D1561863

FINITE FIELDS FOR COMPUTER SCIENTISTS AND ENGINEERS

THE KLUWER INTERNATIONAL SERIES IN ENGINEERING AND COMPUTER SCIENCE

INFORMATION THEORY

Consulting Editor

Robert G. Gallager

FINITE FIELDS FOR COMPUTER SCIENTISTS AND ENGINEERS

by

Robert J. McEliece
California Institute of Technology

KLUWER ACADEMIC PUBLISHERS
Boston / Dordrecht / Lancaster

Distributors for North America:
Kluwer Academic Publishers
101 Philip Drive
Assinippi Park
Norwell, Massachusetts 02061, USA

Distributors for the UK and Ireland:
Kluwer Academic Publishers
MTP Press Limited
Falcon House, Queen Square
Lancaster LA1 1RN, UNITED KINGDOM

Distributors for all other countries:
Kluwer Academic Publishers Group
Distribution Centre
Post Office Box 322
3300 AH Dordrecht, THE NETHERLANDS

Library of Congress Cataloging-in-Publication Data

McEliece, Robert J.
Finite fields for computer scientists and engineers.

(The Kluwer international series in engineering and computer science ; 23)
Bibliography: p.
Includes index.
1. Finite fields (Algebra) I. Title. II. Series: Kluwer international series in engineering and computer science ; SECS 23.
QA247.3.M37 1987 512'.3 86-21145
ISBN 0-89838-191-6

Printed in the United States of America

to my friend
Gus Solomon
who taught me all this stuff

Contents

Preface

This book developed from a course on finite fields I gave at the University of Illinois at Urbana-Champaign in the Spring semester of 1979. The course was taught at the request of an exceptional group of graduate students (including Anselm Blumer, Fred Garber, Evaggelos Geraniotis, Jim Lehnert, Wayne Stark, and Mark Wallace) who had just taken a course on coding theory from me. The theory of finite fields is the mathematical foundation of algebraic coding theory, but in coding theory courses there is never much time to give more than a "Volkswagen" treatment of them. But my 1979 students wanted a "Cadillac" treatment, and this book differs very little from the course I gave in response. Since 1979 I have used a subset of my course notes (corresponding roughly to Chapters 1–6) as the text for my "Volkswagen" treatment of finite fields whenever I teach coding theory. There is, ironically, no coding theory anywhere in the book!

If this book had a longer title it would be "Finite fields, mostly of characteristic 2, for engineering and computer science applications." It certainly does not pretend to cover the general theory of finite fields in the profound depth that the recent book of Lidl and Neidereitter (see the Bibliography) does. What it does do, however, is to give a thorough discussion of the elementary things like what finite fields are, how they are constructed, and how to make computations. (This in Chapters 1–6.) In the final five chapters,

I treat in some depth several topics which are closely related to coding theory but which are rarely covered in the classroom. These topics include two of Elwyn Berlekamp's brilliant recent contributions to the subject, viz., his polynomial factorization algorithm (Chapter 7) and his bit-serial multiplication circuits (Chapter 8). Also, the last three Chapters (9, 10, and 11) include (among other things) what I hope is a "Cadillac" treatment of the theory of m-sequences, an old topic which has recently assumed increased practical importance because of its applications to spread-spectrum communications.

No book is written in a vacuum, least of all this one, and I am happy to acknowledge my debts. I already mentioned the graduate students who forced me to offer a course on finite fields in 1979. Many later students at the University of Illinois and Caltech have criticised and thereby improved the notes. Of these, I would particularly like to thank Doug Whiting, who convinced me of the importance of dual bases and bit-serial arithmetic. (This is in fact the only topic included in the book which was not covered in the 1979 course.) Carl Harris of Kluwer convinced me that the course notes should be published, and has gently but firmly kept me more-or-less on schedule as the manuscript was being prepared. Joanne Clark typed and retyped the manuscript, using Don Knuth's brilliant but often aggravating TEX computer typesetting program. The final preparation of the manuscript was done by Caltech's infallible TEX guru, Calvin Jackson, and I feel the appearance of the book fully justifies Calvin's careful and expert hard work.

And finally I wish to thank Gus Solomon, who first taught me about finite fields, and many other things, quite a number of years ago. Gus has forgotten more about those subjects than I will ever know. Thank you, Gus.

FINITE FIELDS FOR COMPUTER SCIENTISTS AND ENGINEERS

Chapter 1

Prologue

We begin at the beginning. A *field* is a place where you can add, subtract, multiply, and divide. More formally, it is a set F, together with two binary operations, "+" and "·", such that:

1. F is an Abelian group under "+", with identity element 0.
2. The nonzero elements of F form an Abelian group under "·".
3. The distributive law $a \cdot (b + c) = a \cdot b + a \cdot c$ holds.

A field is called *finite* or *infinite* according to whether the underlying set is finite or infinite. Familiar examples of infinite fields include the real numbers, the rational numbers, the complex numbers, and rational functions over a field. We find infinite fields uninteresting. However, we find the following *finite* field extremely interesting:

$$Z_p = \{0, 1, \ldots, p-1\}, \qquad \text{arithmetic mod } p,$$

where p is a prime. It is not obvious that Z_p as defined above is indeed a field, and we shall give a proof in Chapter 4. For now just notice that Z_4 (arithmetic mod 4) is *not* a field, since e.g. 2 has no inverse, i.e., there is no element x such that $2x \equiv 1 \pmod 4$.

However, there is a field with four elements. If we denote its elements as $\{0, 1, 2, 3\}$, the addition and multiplication tables are as follows:

	0	1	2	3
0	0	1	2	3
1	1	0	3	2
2	2	3	0	1
3	3	2	1	0

$+$

	0	1	2	3
0	0	0	0	0
1	0	1	2	3
2	0	2	3	1
3	0	3	1	2

$\cdot$

Soon we will see why this peculiar-looking construction works; at present, the doubtful reader can just verify that it satisfies the axioms! There are many more finite fields, and in this book we will study them all. However, before we can go any further, we'll have to learn some basic facts from algebra. We will begin in Chapter 2 with a study of Euclidean domains and Euclid's algorithm.

Problems for Chapter 1.

1. Show that arithmetic (mod 2), and arithmetic (mod 3), are both fields.

2. Verify that the addition and multiplication tables given in the text do make the set $\{0, 1, 2, 3\}$ into a field.

3. In the four element field described in the text, how many solutions does the polynomial equation $x^3 = 1$ have?

4. Is there a field with just one element?

Chapter 2

Euclidean Domains and Euclid's Algorithm

In 300 B.C. Euclid gave a remarkably simple procedure for finding the greatest common divisor of two integers. Since that time "Euclid's Algorithm" has evolved to become one of the most useful tools in mathematics. In this chapter we will give a fairly complete treatment of Euclid's algorithm.

The modern versions of Euclid's algorithm are techniques for finding greatest common divisors (gcd's), not only of pairs of integers, but also for pairs of polynomials, or indeed for any pair of elements taken from a *Euclidean domain.* The definition of a Euclidean domain follows.

An *integral domain* is a set D, together with two binary operations, $+$ and $\cdot$, such that:

1. The elements of D form an abelian group under $+$; the additive identity element is denoted by 0.
2. The multiplication is associative and commutative, and has an identity element, denoted by 1.
3. The cancellation law holds. That is, if $ab = ac$ and $a \neq 0$, then $b = c$.
4. The distributive law holds. That is, if a, b, and c belong to D, then $a(b+c) = ab + ac$.

A *Euclidean domain* is an integral domain with an added feature: a notion of "size" among its elements. The "size" of $a \neq 0$, denoted $g(a)$, is a nonnegative integer* such that

$$g(a) \leq g(ab) \quad \text{if } b \neq 0; \tag{2.1}$$

and

(2.2) For all $a, b \neq 0$, there exist q and r ("quotient" and "remainder") such that $a = qb + r$, with $r = 0$ or $g(r) < g(b)$.

Here are some examples of Euclidean domains:

- The integers with $g(n) = |n|$.
- Polynomials over a field, with $g(f(x)) = \text{degree } (f)$.
- The Gaussian integers: $\{a + b\sqrt{-1} : \quad a, b \text{ integers}\}$, with $g(a + bi) = a^2 + b^2$.

In the first two examples, properties (2.1) and (2.2) follow trivially—here q and r *are* the quotient and remainder if a is divided by b. The verification of property (2.2) for the Gaussian integers is more interesting however, and is left as an exercise.

Now that we have defined integral domains and Euclidean domains, we'll go on to the main concern of this section, *greatest common divisors.*

In a given integral domain D, an element $a \neq 0$ is said to *divide* another element b, in symbols $a \mid b$, if there is a third element c such that $b = c \cdot a$. If $a \mid b_i$, $i = 1, 2, \ldots, n$, a is said to be a *common divisor* of the b_i's. Finally, if d is a common divisor of $\{b_1, \ldots, b_n\}$, and if every other common divisor of $\{b_1, \ldots, b_n\}$ divides d, d is said to be a *greatest common divisor* of $\{b_1, b_2, \ldots, b_n\}$. We denote this by writing $d = \gcd(b_1, \ldots, b_n)$. We are interested in finding gcd's efficiently; and, if possible, in being able to express the gcd as a linear combination of the b_i's.

Example 2.1. $D =$ the integers, $b_1 = 84$, $b_2 = -140$, $b_3 = 210$. Common divisors include $1, -2, 7$; but the gcd is ± 14; i.e., $\gcd(84, 140, 210) = 14$. Note also that $d = 14 = 84 + 140 - 210$ is a *linear combination* of b_1, b_2, b_3 (see Theorem 2.1, below). ■

---

* If necessary, we will take $g(0) = -\infty$.

Example 2.2. $D = F[x, y]$, the ring of polynomials in two indeterminates over a field F. (Note: D is *not* a Euclidean domain.) Then $\gcd(x, y) = 1$. But notice that 1 cannot possibly be written as a linear combination of x and y, because if $P(x, y)x + Q(x, y)y = 1$, a contradiction arises if we set $x = y = 0$. ∎

Example 2.3. $D = Z[\sqrt{-5}\,]$, the set of complex numbers of the form $a + b\sqrt{-5}$, where a and b are integers. (Note: D is not a Euclidean domain.) Let $b_1 = 9$, $b_2 = 6 + 3\sqrt{-5}$, $d_1 = 3$, $d_2 = 2 + \sqrt{-5}$. Then $d_i \mid b_j$ for $i, j = 1, 2$, as the reader may verify, so d_1 and d_2 are both common divisors of b_1 and b_2. But it is possible to show that there is no number d which simultaneously divides b_1 and b_2 and is divisible by d_1 and d_2. Hence b_1 and b_2 have no greatest common divisor. ∎

We now show that the pathologies present in Examples 2.2 and 2.3 disappear in Euclidean domains.

Theorem 2.1. *If $B = \{b_1, b_2, \ldots, b_n\}$ is any finite subset of a Euclidean domain D, then B has a gcd d, which can be expressed as a linear combination $\sum \lambda_k b_k$ of the b_k's.*

Proof: Let $S = \{\sum_{k=1}^{n} \mu_k b_k : \mu_k \in D\}$, and let d be a nonzero element of S for which $g(d)$ is as small as possible. As an element of S, d can plainly be expressed as a linear combination of the b_k's. We claim that d is a gcd of the b_k's.

First we show that $d \mid b_i$, $i = 1, 2, \ldots, n$. Since $d \neq 0$, by property (2.2), we can write $b_i = q_i d + r_i$, with either $r_i = 0$ or $g(r_i) < g(d)$. Clearly $r_i = b_i - q_i d$ is an element of S. Since d was chosen to have $g(d)$ as small as possible among the *nonzero* elements, this means that $r_i = 0$, i.e., $b_i = q_i d$. Thus d is a common divisor of the b_i's.

Now if c is any other common divisor of the b_i's, say $b_i = q_i' e$, $i = 1, 2, \ldots, n$, since we know that d is a linear combination of the b_i's, $d = \sum \lambda_i b_i = e \sum \lambda_i q_i'$ is a multiple of e. Thus d is as claimed a gcd for $(b_1, \ldots, b_n)$. ∎

Theorem 2.1 assures us that gcd's exist, but it is not very helpful if we actually need to calculate a gcd. Euclid's algorithm will remedy this, but we first need a simple result upon which everything depends.

Theorem 2.2. $\gcd(s,t) = \gcd(s, t - rs)$ *for any elements* s, t, r.

Proof: If d divides both s and t, d certainly divides $t - rs$; so every common divisor of s and t is also a common divisor of s and $t - rs$. Similarly a common divisor of s and $t - rs$ must also be a common divisor of s and $t = (t - rs) + r \cdot s$. Hence the set of common divisors of s and t is the same as the set of common divisors of s and $t - rs$; from this the theorem follows. ∎

Theorem 2.2 is the inductive basis for Euclid's algorithm for integers. The idea is that since $\gcd(s,t) = \gcd(s, t - s)$, one can find the gcd of two integers by continually subtracting the smaller from the larger, until the two numbers are equal, at which point the common value is the gcd. This is exactly how Euclid described his algorithm. The usual modern version is essentially the same, except that repeated subtraction of the same number is replaced by a single division. However, it is important to realize that only subtraction is needed to compute integer gcd's. This means that Euclid's algorithm could easily be taught to junior high school students!

Before proceeding, we will give a somewhat technical result about gcd's that will be referred to many times later on. It gives a simple rule for computing the gcd of two elements of the form $t^n - 1$ and $t^m - 1$.

Theorem 2.3. *Let t be an element in any domain where gcd's exist. Then if m and n are positive integers,*

$$\gcd(t^n - 1, t^m - 1) = t^{\gcd(n,m)} - 1.$$

Proof: Induction on $\max(n, m)$. If $\max(m, n) = 1$, or if $n = m$, the result is trivial. Otherwise we assume $m < n$ and note that $(t^n - 1) - t^{n-m}(t^m - 1) = t^{n-m} - 1$. Hence:

$$\begin{aligned} \gcd(t^n - 1, t^m - 1) &= \gcd(t^m - 1, t^{n-m} - 1) && \text{(Theorem 2.2)} \\ &= t^{\gcd(m,n-m)} - 1 && \text{(induction)} \\ &= t^{\gcd(n,m)} - 1 && \text{(Theorem 2.2 again).} \end{aligned}$$

∎

Corollary 2.4. *Under the same assumptions,*

$$\gcd(x^{q^n} - x, x^{q^d} - x) = x^{q^{\gcd(n,d)}} - x.$$

Proof: Left as an exercise. ∎

Theorem 2.3 is very powerful; for example, it says that $\gcd(2^{15}-1, 2^{20}-1) = 2^5 - 1 = 31$, a fact which would not have been obvious if we had written $2^{15} - 1 = 32767$, $2^{20} - 1 = 1048575$. Similarly, using Theorem 2.3 we can calculate many *polynomial* gcd's effortlessly:

$$\gcd(x^{15} - 1, x^{20} - 1) = x^5 - 1$$

$$\gcd((x+y)^5 - 1, (x+y)^{20} - 1) = (x+y)^5 - 1, \quad \text{etc.}$$

A Curious Fact. If we define the sequence $M_n = 2^n - 1$, Theorem 2.3 tells us that $\gcd(M_n, M_m) = M_{\gcd(n,m)}$. Furthermore, M_n satisfies the recursion $M_0 = 0, M_1 = 1, M_{n+1} = 3M_n - 2M_{n-1}$. It is known that the Fibonacci numbers $F_0 = 0, F_1 = 1, F_{n+1} = F_n + F_{n-1}$ also satisfy $\gcd(F_n, F_m) = F_{\gcd(n,m)}$. More generally, if G_n is a sequence satisfying $G_0 = 0, G_1 = 1$, and for $n \geq 1$, $G_{n+1} = aG_n + bG_{n-1}$, where $\gcd(a, b) = 1$, then $\gcd(G_n, G_m) = G_{\gcd(n,m)}$. (See Problem 11.)

We can finally state Euclid's algorithm, which is a recursive procedure for finding gcd's in any Euclidean domain D.

Suppose we are given two elements $a, b \in D$, both nonzero, and want to find $d = \gcd(a, b)$. For definiteness we assume $g(a) \geq g(b)$. Here is Euclid's algorithm. Define $r_{-1} = a$, $r_0 = b$, and if r_{i-1} is not zero, define r_i recursively by using property (2.2):

$$(2.3) \qquad r_{i-2} = q_i r_{i-1} + r_i, \qquad g(r_i) > g(r_{i-1}).$$

That is, r_i is the "remainder" obtained when r_{i-2} is divided by r_{i-1}, and q_i is the "quotient."

The procedure (2.3) is continued until a remainder of 0 is reached, say $r_{n+1} = 0$ (this must happen, since $g(r_i)$ is strictly decreasing). Then $r_n = \gcd(a, b)$.

To see why this is so, simply note that from (2.3)

$$r_i = r_{i-2} - q_i r_{i-1}, \tag{2.4}$$

and hence by Theorem 2.2, $\gcd(r_{i-2}, r_{i-1}) = \gcd(r_{i-1}, r_i)$. Thus $\gcd(a, b) = \gcd(r_{-1}, r_0) = \cdots = \gcd(r_n, r_{n+1}) = \gcd(r_n, 0) = r_n$.

Example 2.4. Let us use Euclid's algorithm to find gcd(84, 54) over the integers. We begin with $r_{-1} = 84, r_0 = 54$, and proceed as follows:

$$\begin{aligned}
84 &= 1 \cdot 54 + 30; \quad q_1 = 1, \quad r_1 = 30 \\
54 &= 1 \cdot 30 + 24; \quad q_2 = 1, \quad r_2 = 24 \\
30 &= 1 \cdot 24 + 6; \quad q_3 = 1, \quad r_3 = 6 \\
24 &= 4 \cdot 6 + 0; \quad q_4 = 4, \quad r_4 = 0.
\end{aligned}$$

Since $r_4 = 0$, we know that $\gcd(84, 54) = r_3 = 6$. Furthermore, we can "unravel" the above equations to express the gcd 6 as a linear combination of 84 and 54:

$$\begin{aligned}
6 &= 30 - 1 \cdot 24 \\
&= 30 - 1 \cdot (54 - 1 \cdot 30) \\
&= 2 \cdot 30 - 1 \cdot 54 \\
&= 2 \cdot (84 - 54) - 54 \\
&= 2 \cdot 84 - 3 \cdot 54
\end{aligned}$$

∎

The technique of working backwards through the equations $r_{i-2} = q_i r_{i-1} + r_i$ will always produce an expression of the gcd as a linear combination of the original two numbers, though it is a bit tedious and requires us to remember all of the r_i's and q_i's. In the "extended" version of Euclid's algorithm to be presented below, we will see a much better way of expressing the gcd as a linear combination of the original two numbers.

Remark: There is no guarantee that the q_i, r_i in Eq. (2.3) are unique, although in general there will be a fairly obvious choice. For example, consider the following alternative computation of $(84, 54)$, where we again begin with $r_{-1} = 84, r_0 = 54$.

$$
\begin{aligned}
84 &= 2 \cdot 54 - 24; \quad q_1 = 2, \quad r_1 = -24 \\
54 &= -2 \cdot (-24) + 6; \quad q_2 = -2, \quad r_2 = 6 \\
-24 &= -4 \cdot 6 + 0; \quad q_3 = -4, \quad r_3 = 0.
\end{aligned}
$$

Here again we conclude $(84, 54) = 6$; but by using negative remainders, we have reduced the work by one step.

We now give the "extended" version of Euclid's algorithm. It involves as before the sequences $\{r_i\}$ and $\{q_i\}$, defined just as before. It also involves two new sequences $\{s_i\}, \{t_i\}$, defined as follows:

$$
\begin{aligned}
\text{initial conditions: } & s_{-1} = 1, \quad s_0 = 0 \\
& t_{-1} = 0, \quad t_0 = 1 \\
\text{recursion: } & t_i = t_{i-2} - q_i t_{i-1} \\
& s_i = s_{i-2} - q_i s_{i-1}.
\end{aligned}
$$

Again, the algorithm proceeds until $r_{n+1} = 0$, at which point $r_n = \gcd(a, b)$. But now it will also follow that

$$s_n a + t_n b = r_n. \tag{2.5}$$

This is the desired expression for $\gcd(a, b)$ as a linear combination of a and b. To prove that (2.5) is true, we will actually prove a stronger result, viz.

$$s_i a + t_i b = r_i, \tag{2.6}$$

all $i = -1, 0, 1, \ldots, n+1$. We can prove (2.6) by induction. Eq. (2.6) is true for $i = -1, 0$, by the definition of the initial conditions. Assuming Eq. (2.6) is true for $i-2$ and $i-1$, it then follows for i upon substituting Eq. (2.4) for r_i, and the recursive definitions for s_i and t_i in terms of s_{i-1}, s_{i-2}, r_{i-1}, and r_{i-2}.

Example 2.5. Let us now illustrate the use of the extended version of Euclid's algorithm on the numbers of Example 2.1. here is a little table that summarizes the work:

i	s_i	t_i	r_i	q_i
−1	1	0	84	–
0	0	1	54	–
1	1	−1	30	1
2	−1	2	24	1
3	2	−3	6	1
4	−9	14	0	4

Note that the $i = 3$ line shows us that $2 \cdot 84 - 3 \cdot 54 = 6$, as before. (Note also that (2.6) holds for the last line, too. With $i = 4$ we have $9 \cdot 84 - 14 \cdot 54 = 0$.) ∎

Problems for Chapter 2.

1. Show that if an integral domain has more than one element, then $0 \neq 1$.

2. This problem deals with the Gaussian integers.

a. Show that the function g defined by $g(x+iy) = x^2 + y^2$ satisfies properties (2.1) and (2.2).

b. Illustrate your proof in part (a) by finding q and r if $a = 3 + 4i$ and $b = 7 - i$.

c. Use Euclid's algorithm to find an equation of the form $d = sa + tb$, where a and b are as in part (b), and $d = \gcd(a, b)$.

3. Find an equation $\gcd(a, b) = sa + tb$, where:

a. $a = 6711, b = 831$, over the integers.

b. $a = x^8, b = x^6 + x^4 + x^2 + x + 1$, polynomials with coefficients in the field $\{0, 1\}$, with mod 2 arithmetic.

c. $a = 3 + 4i$, $b = 2 + 3i$ (Gaussian integers).

4. The *Fibonacci numbers* are a sequence of integers define as follows. $F_0 = 0$, $F_1 = 1$, and for $n \geq 2$, $F_n = F_{n-1} + F_{n-2}$. Experiment with Euclid's algorithm as applied to consecutive Fibonacci numbers F_n and F_{n+1} and answer the following:

a. What is $\gcd(F_{n+1}, F_n)$?

b. In the computation of $\gcd(F_{n+1}, F_n)$, what are the q_i's, r_i's, s_i's, and the t_i's?

c. Express $\gcd(F_{n+1}, F_n)$ as an *explicit* linear combination of F_n and F_{n+1}.

5. We have seen that, given a pair of elements a and b in a Euclidean domain, Euclid's algorithm can be used to produce another pair of elements s and t such that $as + bt = \gcd(a, b)$. But in general there will be many such pairs. For example, $\gcd(6, 4) = 2$ and $6 \cdot 1 + 4 \cdot (-1) = 6 \cdot (-1) + 4 \cdot 2 = 2$. Here is the problem. Given just one pair (s_0, t_0) with $as_0 + bt_0 = \gcd(a, b)$, describe a procedure for finding all such pairs. Illustrate your results in the case $a = 6$, $b = 4$.

6. Show that the integral domains described in Examples 2.3 and 2.3 are not Euclidean domains. [Hint: Use Theorem 2.1.]

7. Prove Corollary 2.4.

8. Prove the following properties of Euclid's algorithm.

a. $t_i r_{i-1} - t_{i-1} r_i = (-1)^i a$.

b. $s_i r_{i-1} - s_{i-1} r_i = (-1)^i b$.

c. $s_i t_{i-1} - s_{i-1} t_i = (-1)^{i+1}$.

d. $s_i a + t_i b = r_i$.

9. If Euclid's algorithm is applied to a pair of polynomials $a(x)$ and $b(x)$, show the following:

a. $\deg(s_i) + \deg(r_{i-1}) = \deg(b)$.

b. $\deg(t_i) + \deg(r_{i-1}) = \deg(a)$.

10. Show that the set of numbers of the form $a + b\sqrt{2}$, where a and b are integers, forms a Euclidean domain with $g(a + b\sqrt{2}) = a^2 + 2b^2$.

11. Let $G_0, G_1, \ldots$ be a sequence of integers defined by $G_0 = 0$, $G_1 = 1$, and for $n \geq 2$, $G_n = aG_{n-1} + bG_{n-2}$, where a and b are integers with $\gcd(a, b) = 1$.

a. Prove that $\gcd(G_n, b) = 1$, for $n \geq 1$.

b. Prove that $\gcd(G_n, G_{n-1}) = 1$, if $n \geq 2$.

c. Prove that for $m \geq 1$, and $n \geq m + 1$,

$$G_n = G_{m+1}G_{n-m} + bG_mG_{n-m-1}.$$

d. Prove that if $n \geq m$, then $\gcd(G_n, G_m) = \gcd(G_m, G_{n-m})$.

e. Finally prove that $\gcd(G_n, G_m) = G_{\gcd(n,m)}$.

Chapter 3

Unique Factorization in Euclidean Domains

In this chapter we will prove the unsurprising but very important fact that factorization into primes is unique in a Euclidean domain. We start with a few definitions.

Let D be an arbitrary Euclidean domain. A *unit* $u \in D$ is any divisor of 1; i.e., u is a unit iff there exists $v \in D$ such that $uv = 1$. Examples: If $D =$ the ordinary integers, the units are ± 1; if $D =$ polynomials over a field k, the units are the scalars, i.e., the polynomials of degree 0; if $D =$ the Gaussian integers, the units are $\pm 1, \pm i$.

Two elements $a, b \in D$ are called *associates* if $a = ub$ for some unit u. The relation of association is easily seen to be an equivalence relation. Examples: $+3$ and -3 are associate integers; $x^2 + 2x + 9$ is associate to $5x^2 + 10x + 45$; $1 + i$ is associate to $1 - i$.

If $b \in D$, a *factorization* of b is any expression of the form $b = a_1 a_2 \cdots a_r$. For example, if $uv = 1$ (i.e., if u and v are units), then $b = b \cdot u \cdot v$. This is, of course, trivial and more generally we say that the factorization $b = a_1 \cdots a_r$ is *trivial factorization* if each of the a_i's is either a unit or an associate of b. (Of course, only one of the a_i's can be an associate of b.)

An element $p \in D$ which has the property that every possible factorization of p is trivial is called a *prime*. The only exception to this is that we do not consider units themselves to be primes. If we agree to call a divisor

d of b a *proper divisor* if d is not an associate of b, it follows that the only proper divisors of primes are units.

Example 3.1. In the ordinary integers, the primes are, of course, ± 2, ± 3, ± 5, ± 7, $\pm 11, \ldots$. ∎

Example 3.2. In the case of polynomials, the term *prime* is usually replaced by the term *irreducible*. We will have a lot to say about irreducible polynomials in future chapters. ∎

Example 3.3. In the Gaussian integers, it is known that the primes are the ordinary rational primes congruent to 3 (mod 4), i.e., $3, 7, 11, 19, 23, \ldots$, and associates, and complex primes of the form $a + b\sqrt{-1}$, whose absolute value $a^2 + b^2$ is either equal to 2 or to a rational prime congruent to 1 (mod 4), i.e., $1+i$, $2+i$, $3+2i$, $4+i$, $5+2i$, $6+i, \ldots$, and associates. ∎

In any Euclidean domain, two elements a and b are said to be *relatively prime* if their gcd is 1 (or any other unit). It follows then from Theorem 2.1 that it is possible to express 1 as a linear combination of a and b:

Lemma 3.1. *If a and b are relatively prime, then there exist s and t such that $as + bt = 1$.* ∎

Lemma 3.2. *If $p \in D$ is prime, and if $p \not| \ a$ (read "p does not divide a") then p and a are relatively prime.*

Proof: Let d be a common divisor of p and a. Since p is prime, d must either be a unit or an associate of p. Since however $p \not| \ a$, no associate of p can divide a, and we conclude that d is a unit. Thus any common divisor of p and a is a unit, and so $\gcd(p, a) = 1$. ∎

Lemma 3.3. *If $p \not| \ a$ then there exist elements $s, t \in D$, such that $ps + at = 1$.*

Proof: Follows from Lemmas 3.1 and 3.2. ∎

Lemma 3.4. *If p is a prime and $p \mid ab$, then either $p \mid a$ or $p \mid b$ (or both).*

Proof: If $p \mid a$, there is nothing to prove. Otherwise $p \nmid a$ and so by Lemma 3.3 there exists s and t such that $ps + at = 1$. Multiplying this by b, we see that $b = pbs + abt$. Since $p \mid ab$, this shows that $p \mid b$. ∎

Lemma 3.5. *In a Euclidean domain, if a is a proper divisor of b, then $g(a) < g(b)$.*

Proof: Suppose $b = ac$, with c not a unit. (This is equivalent to saying that a is not an associate of b.) We divide a by b, using property (2.2):

$$a = qb + r, \quad g(r) < g(b).$$

Then $r = a - qb = a - qac = a(1 - qc)$. Now since c is not a unit $1 - qc \neq 0$ and so by property (2.1), it follows that $g(r) \geq g(a)$. Hence $g(a) \leq g(r) < g(b)$, as claimed. ∎

We are now in a position to state and prove the unique factorization theorem for Euclidean domains.

Theorem 3.6. *Let $b \in D$, not a unit. Then (i) b can be written as a product of primes*

$$b = p_1 p_2 \cdots p_r, \qquad \text{each } p_i \text{ a prime; and}$$

(ii) If b is written in another way as a product of primes, say

$$b = q_1 q_2 \cdots q_s,$$

then $r = s$, and after a suitable renumbering, p_i and q_i are associates, $i = 1, 2, \ldots, s$.

Proof: (i) Induction on $g(b)$. If b itself is prime, the expression $b = b$ satisfies us. Otherwise b has a nontrivial factorization $b = a \cdot c$ where both a and c are proper divisors of b. By Lemma 3.5 $g(a) < g(b)$, $g(c) < g(b)$, and

so by induction each of a and c can be expressed as a product of primes: $a = p_1 p_2 \cdots p_j$, $c = p_{j+1} \cdots p_r$. Hence, $b = ac = p_1 \cdots p_r$, as asserted.
(ii) Now suppose b has two such expressions, viz.,

$$\begin{aligned} b &= p_1 p_2 \cdots p_r \\ &= q_1 q_2 \cdots q_s. \end{aligned}$$

Then $p_1 \mid q_1 q_2 \cdots q_s$ and so by an obvious extension of Lemma 3.4, p_1 must divide one of the q_i's. By renumbering we assume, in fact, that $p_1 \mid q_1$. Since, however, q_1 is prime and p_1 is not a unit, it follows that p_1 and q_1 are associates, i.e., $q_1 = \epsilon_1 p_1$ for some unit ϵ_1. Hence $p_1 p_2 \cdots p_r = (\epsilon_1 p_1) q_2 \cdots q_s$ and so $p_2 p_3 \cdots p_r = q_2' q_3 \cdots q_r$, with $q_2' = \epsilon_1 q_2$. This last equation is an expression for $b' = b/p_1$ (a proper divisor of b) as a product of primes in two different ways; by induction on $g(b)$ we now conclude that $r = s$ and p_i and q_i are associates for $i = 2, \ldots, r$. ■

Now that we know factorization is unique, we can describe a conceptually simple but computationally clumsy method for computing gcd's. If $b \in D$, and if p is prime, we define

$$e_p(b) = \max\{e : p^e \mid b\}. \tag{3.1}$$

In words, $e_p(b)$ is the highest power of the prime p that divides b. If we know $e_p(b)$ for all primes p, we will be able to compute b except for unit factors. For example, if we consider the integers and are given $e_2(b) = 2$, $e_3(b) = e_5(b) = 1$, $e_p(b) = 0$ for all $p \geq 7$, then clearly $b = \pm 2^2 \cdot 3^1 \cdot 5^1 = \pm 60$. Furthermore, we can compute the gcd of two numbers (up to association) from the e_p–function of those numbers. Omitting the details, which the reader is encouraged to supply, we have, if $d = \gcd(a, b)$, that

$$e_p(d) = \min\{e_p(a), e_p(b)\}. \tag{3.2}$$

We could have started with (3.2) as the definition of gcd's, and, in fact, something like this is often taught to junior high school students. But to actually compute gcd's using (3.2), one would have to compute the prime factorization of a and b first, and factoring large integers isn't easy. In fact, if a and b were say 100-digit integers, no known factorization algorithm would succeed in finding the needed prime factorizations, even if a high-speed computer were used for many hours. But it's not necessary to factor a and b in order to find their gcd! Euclid's algorithm somehow avoids the difficulties inherent in factorization and goes directly to the gcd; if a and b were 100-digit integers, a program written by a competent programmer would succeed in finding $d = \gcd(a, b)$ in at most a few seconds.

Problems for Chapter 3.

1. Let D be the integral domain consisting of all polynomials with coefficients in a field F. What are the units?

2. Show that the only Gaussian integer units are ± 1 and $\pm i$.

3. Show that the relation of *associate* is an equivalence relation, i.e. that it is reflexive, symmetric, and transitive.

4. In a Euclidean domain, show that ϵ is a unit if and only if $g(\epsilon) = g(1)$.

5. Show that in a trivial factorization $b = a_1 a_2 \cdots a_r$ in a Euclidean domain of a nonunit b, exactly one of the factors a_i is an associate of b.

6. We would like to ask you to verify the assertion made in the text about the Gaussian integer primes, but this would be too difficult. Instead, we ask the following:
   a. Show that $1 + i$, $2 + i$, 3, $3 + 2i$, and 7 are prime.
   b. Show that 5, 13, and 17 are not prime.

7. Show that (3.2) is valid.

8. Factor the following Gaussian integers: 8, $8 + 4i$, 10, 12, $45 + 3i$, 60.

9. Let D be the integral domain $F_2[x]$, consisting of all polynomials with coefficients in the field F_2 (integers mod 2).

a. What are the units?

b. Factor the following polynomials into their irreducible factors: $x^2 + 1$, $x^2 + x + 1$, $x^2 + x$.

10. If p is a prime and $p \mid a_1 a_2 \cdots a_n$, show that p divides a_i for some i.

11. Consider the integral domain D consisting of the numbers of the form $a + b\sqrt{-3}$, where a and b are integers.

a. Show that the number 4 has two essentially different factorizations into prime factors:

$$4 = 2 \cdot 2 = (1 + \sqrt{-3})(1 - \sqrt{-3}).$$

b. Show that D is not a Euclidean Domain.

Chapter 4

Building Fields from Euclidean Domains

In this chapter we will show that given a Euclidean domain D and a prime $p \in D$, it is possible to construct a field (which may or may not be finite), called "$D \bmod p$."

We begin by considering an arbitrary but fixed element $m \in D$, not necessarily prime, and define an equivalence relation "$\sim$": $a \sim b$ if and only if $m \mid (a - b)$. One easily verifies that this relation is indeed an equivalence relation, i.e., it is reflexive, symmetric, and transitive. It is usual to represent this particular relation not by "$\sim$", but rather by the famous notation introduced by Gauss:

$$(4.1) \qquad a \equiv b \pmod{m} \quad \text{iff} \quad m \mid (a - b).$$

This equivalence relation, like any equivalence relation, partitions its underlying set (in this case D) into a certain number of disjoint subsets called *equivalence classes.* If $a \in D$, we will denote the unique equivalence class containing a by $\bar{a}$. Our plan is to define an arithmetic (addition and multiplication) on these equivalence classes.

Example 4.1. Let D be the integers, $m = 4$. Then there are four equivalence classes:

$$\overline{0} = \{0, \pm 4, \pm 8, \pm 12, \ldots\}$$
$$\overline{1} = \{\ldots, -7, -3, 1, 5, 9, 13, \ldots\}$$
$$\overline{2} = \{\ldots, -6, -2, 2, 6, 10, 14, \ldots\}$$
$$\overline{3} = \{\ldots, -5, -1, 3, 7, 11, 15, \ldots\}.$$

Notice that the equivalence class labeled $\overline{1}$ could just as easily be called $\overline{5}$, $\overline{-7}$, etc. It is however traditional to represent a particular equivalence class by its smallest nonnegative element, and indeed the notation "$a \bmod m$" is usually taken to mean the least nonnegative element of $\overline{a}$. ∎

We now want to define addition and multiplication of the mod m equivalence classes. For addition, the natural definition is this:

$$\overline{a} + \overline{b} = \overline{a+b}. \tag{4.2}$$

Similarly, we define multiplication as follows:

$$\overline{a} \cdot \overline{b} = \overline{(a \cdot b)}. \tag{4.3}$$

Here a small technical difficulty arises. The definitions (4.2) and (4.3) appear to depend on the particular elements a and b chosen to represent the equivalence classes. To illustrate, let's consider the previous example, and try to add $\overline{1}$ and $\overline{2}$. By (4.2) $1+2=3$ and so $\overline{1}+\overline{2}=\overline{3}$, by (4.2). But $\overline{1} = \overline{-7}$, and $\overline{2} = \overline{14}$, so an equally valid application of (4.2) would yield $\overline{1}+\overline{2} = \overline{-7}+\overline{14} = \overline{-7+14} = \overline{7}$. Fortunately, however, $\overline{3} = \overline{7}$, because $3 - 7 \equiv 0 \pmod 4$, so in this case we appear to have been lucky.

But of course it wasn't luck at all, the same thing always happens. Thus consider the calculation of $\overline{a} + \overline{b}$ in general. By (4.2) we have

$$\overline{a} + \overline{b} = \overline{a+b}.$$

Now suppose $\overline{a_1} = \overline{a}, \overline{b_1} = \overline{b}$, i.e., $a_1 \equiv a \pmod m$, and $b_1 \equiv b \pmod m$. Using the representatives a_1 and b_1 for $\overline{a}$ and $\overline{b}$, we calculate, again using

(4.2),

$$\overline{a} + \overline{b} = \overline{a_1 + b_1}.$$

Question: Is $\overline{a_1 + b_1}$ the same as $\overline{a+b}$? Answer: yes, because this statement is equivalent to saying $a + b \equiv a_1 + b_1 \pmod{m}$, and this follows immediately from $a \equiv a_1 \pmod{m}$ and $b \equiv b_1 \pmod{m}$. A similar argument works for multiplication of equivalence classes, and we conclude that the definitions (4.2) and (4.3) are well-defined.

It is easy to see that the set of equivalence classes forms a *ring* with respect to this arithmetic. The addition identity is given by

$$\overline{0} = \{x \in D : x \equiv 0 \pmod{m}\},$$

and the multiplication identity by

$$\overline{1} = \{x \in D : x \equiv 1 \pmod{m}\},$$

where 0 and 1 denote the additive and multiplicative identities in the Euclidean domain D. This ring is denoted by the symbol $D \bmod m$. For example, if Z denotes the set of integers, then $Z \bmod 4$ is a ring containing 4 elements, as we saw in Example 4.1.

We are interested in knowing whether $D \bmod m$ is a field. This will be the case provided that multiplicative inverses exist, i.e., if for any $\overline{a} \neq \overline{0}$ there exists a $\overline{b}$ such that

$$\overline{a} \cdot \overline{b} = \overline{1}. \tag{4.4}$$

Unfortunately this is not always the case. Example 4.1 gives a typical counterexample. There one easily computes that $\overline{2} \cdot \overline{2} = \overline{0}$, i.e., $Z \bmod 4$ contains zero divisors, and cannot therefore be a field. We come now to the main result of this section.

Theorem 4.1. *If p is prime, $D \bmod p$ is a field.*

Proof: The above discussion gives all the needed properties except the existence of inverses. So let $\overline{a}$ be an element of $D \bmod p$, $\overline{a} \neq \overline{0}$; we must show that there exists an element $\bar{b}$ such that (4.4) holds. Now $\bar{a} \neq 0$ means that $p \not| \ a$. Thus by Lemma 3.3 , there exist elements b and t in D such that $ab + pt = 1$. Thus $ab \equiv 1 \pmod{p}$, and this in turn is equivalent to (4.4). Thus $\bar{b}$ is the inverse of $\bar{a}$; moreover, Euclid's algorithm gives a constructive procedure for finding this inverse. ∎

In the remainder of this section we'll give some examples of the fields $D \bmod p$.

Example 4.2. Let $D = Z$, the integers, $p = 13$. Then $D \bmod p$ has 13 elements, which we may denote by $\overline{0}, \overline{1}, \ldots, \overline{12}$. Then for example

$$\overline{5} \cdot \overline{10} = \overline{11}, \ \overline{4} \cdot \overline{7} = \overline{2}, \quad \text{etc.}$$

Let us find the inverse of $\overline{6}$. We apply Euclid's algorithm to 6 and 13 to find a linear combination of 6 and 13 equal to 1. We find that $6 \cdot 11 - 13 \cdot 5 = 1$. Thus $\overline{6} \cdot \overline{11} = \overline{1}$, i.e., $(\overline{6})^{-1} = \overline{11}$. ∎

Example 4.3. Take $D = Z$ again, but now let p be an arbitrary prime. The resulting important *finite field* has exactly p elements $\{\overline{0}, \overline{1}, \ldots, \overline{p-1}\}$; it is commonly denoted by either of the two symbols F_p or $GF(p)$. This construction yields infinitely many finite fields, since there are infinitely many primes. But there are many other finite fields, as we shall see. ∎

Example 4.4. Here $D = R[x]$, the set of polynomials with real coefficients. We take as p the irreducible polynomial $p = x^2 + 1$. In this case there are infinitely many equivalence classes. But any polynomial $f(x)$ is congruent $(\bmod\, x^2+1)$ to exactly one polynomial of degree 0 or 1: $f(x) \equiv r(x) \pmod{x^2+1}$, where $r(x)$ is the remainder when $f(x)$ is divided by $x^2 + 1$. Thus we can take as representatives of the equivalence classes the polynomials of the form $a + bx$ for $a, b \in R$. Adding two of these equivalence classes is simple:

$$\overline{(a + bx)} + \overline{(a' + b'x)} = \overline{(a + a') + (b + b')x}\,.$$

Multiplication is a little harder. Our first attempt is naturally

$$\overline{(a + bx)}\ \overline{(a' + b'x)} = \overline{aa' + (ab' + a'b)x + bb'x^2}\,. \tag{4.5}$$

But the polynomial on the right side of (4.5) has degree 2 (unless b or b' is zero), and we have selected linear polynomials as equivalence class representatives. We circumvent this problem by noting that $\overline{x^2} = \overline{-1}$, and so actually

$$\overline{(a+bx)(a'+b'x)} = \overline{(aa'-bb')+(ab'+a'b)x} \tag{4.6}$$

is the proper rule for the multiplication. Does this look familiar? It ought to, because if we drop the bars in (4.6) and replace the symbol x with the symbol i, we get

$$(a+bi)(a'+b'i) = (aa'-bb')+(ab'+a'b)i,$$

which is just the ordinary rule for multiplying *complex numbers*. We have just constructed the field of complex numbers! ∎

Example 4.5. $D = R[x]$ as before, but this time, let's take as our prime p an *arbitrary* irreducible quadratic polynomial $p(x) = Ax^2+Bx+C$. Irreducibility over the reals is equivalent to having negative discriminant, i.e., $\Delta = B^2 - 4AC < 0$.

Again we can take as representatives of the (mod p) equivalence classes the polynomials of degree ≤ 1. To multiply two elements of this field, say $(a+bx)$ and $(a'+b'x)$, we proceed as follows:

$$(a+bx)(a'+b'x) = aa' + (ab'+a'b)x + bb'x^2;$$

but

$$x^2 \equiv -(B/A)x - (C/A) \pmod{p(x)},$$

and so

$$(a+bx)(a'+b'x) \equiv (aa'-bb'C/A) + (ab'+a'b-bb'B/A)x \pmod{p(x)}.$$

Thus if we regard $a+bx$ not as a polynomial but as a 2–dimensional vector (a,b), we see that provided $B^2-4AC<0$, the rule

$$(a,b)\cdot(a',b') = (aa'-bb'C/A, ab'+a'b-bb'B/A)$$

makes the space of two dimensional vectors with real components into a field. From a more sophisticated viewpoint we can see that all we've "really" done is to adjoin a complex root of the equation $p(x) = 0$, e.g., $z_0 = (-B + \sqrt{\Delta})/2$, to the field of real numbers. The components of the vector (a, b) are just the coordinates of a certain complex number with respect to the basis $(1, z_0)$ of the complex numbers. In other words, although it *appears* that we've constructed a different field for each possible $p(x)$ with $B^2 - 4AC < 0$, in fact all of these fields are *isomorphic*. The apparent differences are due only to the fact that the rule for multiplying complex numbers is different in different coordinate systems. The moral of the story is that if $D = R[x]$, any irreducible $p(x)$ of degree 2 leads to the same field, viz., the complex numbers. ∎

Example 4.6. Take $D = R[x]$ again, and let $p(x)$ be an irreducible polynomial of degree 3. Using our $D \bmod p$ construction, we will be led to a rule for multiplying 3–dimensional Euclidean vectors, which, when combined with ordinary component-by-component addition, makes a field. But in advanced texts on algebra it is shown that no such multiplication exists! The explanation of this paradox is that there are no irreducible cubics in $R[x]$! Indeed, the "fundamental theorem of algebra" implies that any polynomial in $R[x]$ of any degree factors into a product of real linear polynomials times a product of real quadratic polynomials, the roots of the quadratics occurring in complex conjugate pairs. ∎

Our next example will illustrate the use of the division algorithm in $F_p[x]$.

Example 4.7. In the domain $F_{13}[x]$, let $a(x) = x^8 + x^6 + 10x^4 + 10x^3 + 8x^2 + 2x + 8$, and $b(x) = 3x^6 + 5x^4 + 9x^2 + 4x + 8$. Let's see if we can find two polynomials $q(x)$ and $r(x)$ such that

$$a(x) = q(x)b(x) + r(x); \qquad \deg(r) < \deg(b). \tag{4.7}$$

To do this, we just divide the polynomial $a(x)$ by the polynomial $b(x)$, as is usually done with "synthetic division," only we must remember that the arithmetic is done in F_{13}:

$$
\begin{array}{r|ccccccccc}
 & & & & & & & 9 & 0 & 7 \\
\hline
3\,0\,5\,0\,9\,4\,8 & 1 & 0 & 1 & 0 & 10 & 10 & 8 & 2 & 8 \\
 & 1 & 0 & 6 & 0 & 3 & 10 & 7 & & \\
\cline{4-10}
 & & & 8 & 0 & 7 & 0 & 1 & 2 & 8 \\
 & & & 8 & 0 & 9 & 0 & 11 & 2 & 4 \\
\cline{4-10}
 & & & & & 11 & 0 & 3 & 0 & 4
\end{array}
$$

Thus $q(x) = 9x^2 + 7$, $r(x) = 11x^4 + 3x^2 + 4$. ■

We note that in a general polynomial domain $k[x]$, the "size" function $g(a) = \text{degree}(a)$ enjoys this nice extra property:

$$g(ab) = g(a) + g(b). \tag{4.8}$$

From (4.6), it follows that the $r(x)$ in (4.5) is unique.

In our last example, we will study one case of the $D \bmod p$ construction in detail, when $D = F_p[x]$, i.e., when D is the set of polynomials in the indeterminate x, with coefficients in the finite field $F_p = Z \bmod p$. This particular type of construction is the key to understanding finite fields, for as we will eventually see, every finite field can be constructed this way.

Example 4.8. Let $D = F_2[x]$, $p(x) = x^3 + x + 1$. First note that p is indeed irreducible, because $p(0) = p(1) = 1$, so $p(x)$ has no zeroes in F_2; and a cubic with no zeroes in a field must be irreducible over that field.

Since $p(x)$ is cubic, we can take as representatives of the (mod p) equivalence classes the 8 polynomials of the form $a_2x^2 + a_1x + a_0$, $a_i \in F_2$, which may also be thought of as "generating functions" for the 8 vectors $[a_2, a_1, a_0]$ with coefficients in F_2.

How do we multiply these eight field elements (vectors)? We start by multiplying the corresponding polynomials:

$$
\begin{aligned}
(a_2x^2 + a_1x + a_0)&(b_2x^2 + b_1x + b_0) \\
&= a_2b_2x^4 + (a_2b_1 + a_1b_2)x^3 \\
&+ (a_2b_0 + a_1b_1 + a_0b_2)x^2 \\
&+ (a_1b_0 + a_0b_1)x + a_0b_0.
\end{aligned} \tag{4.9}
$$

This polynomial is of degree 4, and so won't do as a field element (equivalence class representative). We must reduce the quartic and cubic terms. To do this, note that

$$x^3 \equiv x + 1 \pmod{x^3 + x + 1}$$
$$x^4 \equiv x^2 + x \pmod{x^3 + x + 1}.$$

Thus $a_2b_2x^4 \equiv a_2b_2x^2 + a_2b_2x$, and $(a_2b_1 + a_1b_2)x^3 \equiv (a_2b_1 + a_1b_2)x + (a_2b_1 + a_1b_2) \pmod{x^3 + x + 1}$. It follows that the rule for multiplying the vector $[a_2, a_1, a_0]$ by the vector $[b_2, b_1, b_0]$ is $[a_2, a_1, a_0] \cdot [b_2, b_1, b_0] = [c_2, c_1, c_0]$, where:

$$\begin{aligned} c_2 &= a_2b_2 + a_2b_0 + a_0b_2 \\ c_1 &= a_2b_2 + a_2b_1 + a_1b_2 + a_1b_0 + a_0b_1 \\ c_0 &= a_2b_1 + a_1b_2 + a_0b_0. \end{aligned} \tag{4.10}$$

This is not a very simple or transparent rule, but by Theorem 4.1, we know that it does make the three-dimensional vector space over F_2 into a field. Fortunately, however, there is a much simpler way to view this particular field, which we shall temporarily denote by the symbol $GF(8)$.

The first step is to calculate the first few powers of x modulo $p(x) = x^3 + x + 1$:

$$\begin{aligned} x^0 &\equiv 1 \\ x^1 &\equiv x \\ x^2 &\equiv x^2 \\ x^3 &\equiv x + 1 \\ x^4 &\equiv x^2 + x \\ x^5 &\equiv x^3 + x^2 \equiv x^2 + x + 1 \\ x^6 &\equiv x^3 + x^2 + x \equiv x^2 + 1 \\ x^7 &\equiv x^3 + x \equiv 1, \end{aligned}$$

all mod $x^3 + x + 1$. It follows that the sequence of powers of x $\pmod{p(x)}$ is periodic, of period 7. We now return to the field $GF(8)$, and denote the equivalence class $\bar{x}$ by α. As a three-dimensional vector, $\alpha = [0, 1, 0]$. Using the table of the powers of x modulo $p(x)$, we can make the following table of the powers of α.

$$\begin{aligned}
\alpha^0 &= 1 \\
\alpha^1 &= \alpha \\
\alpha^2 &= \alpha^2 \\
\alpha^3 &= \alpha + 1 \\
\alpha^4 &= \alpha^2 + \alpha \\
\alpha^5 &= \alpha^2 + \alpha + 1 \\
\alpha^6 &= \alpha^2 + 1 \\
\alpha^7 &= 1.
\end{aligned}$$

Thus the first 7 powers of α are all distinct in $GF(8)$; but since there are only 7 nonzero elements in $GF(8)$, it follows that every nonzero element of $GF(8)$ is a power of α. So we can use α as a base for "logarithms," if we agree that

$$\log_\alpha(\beta) = k \quad \text{means} \quad \alpha^k = \beta.$$

Here then is a table of logarithms and antilogarithms in $GF(8)$ (we again view the elements of $GF(8)$ as three-dimensional vectors over F_2):

β	$\log_\alpha \beta$	k	α^k
000	$\star$	$\star$	000
001	0	0	001
010	1	1	010
011	3	2	100
100	2	3	011
101	6	4	110
110	4	5	111
111	5	6	101

Thus to compute say $a \cdot b = [110] \cdot [111]$, we could use (4.8) with $a_2 = 1$, $a_1 = 1$, $a_0 = 0$, $b_2 = 1$, $b_1 = 1$, $b_0 = 1$, or we could note that $\log_\alpha(a) = 4$, $\log_\alpha(b) = 5$, so $a \cdot b = \alpha^{4+5} = \alpha^9 = \alpha^2 = [100]$. (Note that exponents on α can be reduced (mod 7), since $\alpha^7 = 1$.) ∎

Problems for Chapter 4.

1. Show that the relation defined by (4.1) is indeed an equivalence relation. What can you say about the special case $m = 0$?

2. Verify that multiplication as defined by (4.3) doesn't depend on the choice of equivalence class representative.

3. Consider the ring $D \bmod m$, defined in the text. Can it ever happen that $\overline{0} = \overline{1}$? If so, is the corresponding ring a field?

4. Show that a ring with zero divisors, i.e., nonzero elements a and b such that $ab = 0$, cannot be a field.

5. In Example 4.4, we noted that $\overline{x^2} = \overline{-1}$. Explain why this is so.

6. Show that (4.8) is true.

7. Use (4.8) to show that the remainder $r(x)$ in (4.7) is unique.

8. In the field $GF(8)$ of Example 4.8, verify that $[110] \cdot [111] = [100]$, using (4.8).

9. Given that $x^4 + x + 1$ is irreducible in $F_2[x]$, construct a field with 16 elements.
   a. Give a rule, analogous to (4.10), for multiplying.
   b. Construct a "logarithm table" like the one in Example 4.8.

10. Let D be the domain of Gaussian integers, and let p be the prime $1+i$. Give a complete description of the field $D \bmod p$.

11. Consider the Euclidean domain $F_3[x]$, i.e., the set of polynomials over the field of integers mod 3. Given that $p(x) = x^2 - x - 1$ is irreducible, construct the field $F_3[x] \bmod p(x)$, by giving the multiplication and addition tables.

Chapter 5

Abstract Properties of Finite Fields

In the last chapter we saw that given a prime field F_p and an mth degree irreducible polynomial $p(x)$ in $F_p[x]$, we can construct a field with p^m elements. Of course we don't yet know whether there are any such polynomials! Fortunately it turns out that such polynomials do exist, and we shall prove that they do in Chapter 6. In this chapter we will prepare for the proof. This preparation will consist of an investigation of some of the important structural properties of finite fields, properties that do not depend on how the field was constructed.

We shall begin by assuming the existence of a finite field F with q elements, and investigate the logical consequences. We already know that for some values of q there are finite fields, e.g. if q is a prime, or if $q = 8$ (Example 4.8), so this investigation isn't vacuous. Along the way, we'll discover many useful facts.

Theorem 5.1. *The number of elements q must be a power of a prime: $q = p^m$, p prime.*

Proof: Let 1 denote the multiplicative identity in F. Define a sequence

$\{u_0, u_1, u_2, \ldots\}$ in F as follows:

$$(5.1) \qquad u_0 = 0, \quad u_n = u_{n-1} + 1, \qquad \text{for } n = 1, 2, \ldots.$$

It follows from this definition that for arbitrary m and n,

$$(5.2a) \qquad u_{m+n} = u_m + u_n$$

$$(5.2b) \qquad u_{mn} = u_m \cdot u_n.$$

Now since F is finite not all the u_n's can be distinct; let $u_k = u_{k+c}$ be the first repeat, i.e., the elements $u_0, u_1, \ldots, u_{k+c-1}$ are all distinct, but $u_{k+c} = u_k$. Then since by (5.2a), $u_{k+c} - u_k = u_c$, it follows that $u_c = 0$, and so 0 is in fact the first element in the sequence $\{u_n\}$ to occur twice, and so the elements $\{u_0, u_1, \ldots, u_{c-1}\}$ are all distinct.

The integer c (necessarily ≥ 2) is called the *characteristic* of the field. We assert that *c must be a prime.* For if on the contrary $c = a \cdot b$ with $1 \leq a \leq c$ and $1 \leq b \leq c$, then it follows from (5.2b) that $u_c = u_a \cdot u_b$. But $u_c = 0$, $u_a \neq 0$, $u_b \neq 0$, and so this is not possible. We conclude that c is, indeed, prime, and from now on we replace the letter c by the letter p to remind ourselves of this fact.

It is clear that the subset $\{u_0, u_1, \ldots, u_{p-1}\}$ of F is a *subfield* of F, since by (5.2) it is closed under $+$ and $\times$. Indeed it is isomorphic to the field $F_p = Z \bmod p = \{0, 1, \ldots, p-1\}$, if we make the obvious identification $u_i \leftrightarrow i$. Thus it is possible to view F as a vector space over F_p. Letting $\{\omega_1, \omega_2, \ldots, \omega_m\}$ denote a basis (necessarily finite) for F over F_p, we see that each element $\alpha \in F$ has a unique expansion of the form

$$(5.3) \qquad \alpha = a_1\omega_1 + a_2\omega_2 + \cdots + a_m\omega_m,$$

where each a_i is an element of F_p. Since there are p possibilities for each a_i, it follows from (5.3) that the field contains exactly p^m elements. ∎

Note on the proof of Theorem 5.1:

- The last part of the proof, where we invoked a little of the theory of vector spaces, can be done directly, as follows:

"We have F_p as a subfield of F. Let $\omega_1 \in F - F_p$; there are p elements in F of the form $a_1\omega_1$, $a_1 \in F_p$. If $q = p$, we are done. Otherwise choose $\omega_2 \in F$ not of the form $a_1\omega_1$. There are p^2 elements in F of the form $a_1\omega_1 + a_2\omega_2$. If $q = p^2$, again, QED. Otherwise choose ω_3 not of the form $a_1\omega_1 + a_2\omega_2$ and consider all sums $a_1\omega_1 + a_2\omega_2 + a_3\omega_3$, and so on."

Theorem 5.1 sheds considerable light on the structure of the additive group of F. It shows that the elements of F can be viewed as m–tuples of elements from F_p, and that if we take this view, then $(a_1, \ldots, a_m)+(b_1, \ldots, b_m) = (a_1+b_1, \ldots, a_m+b_m)$. However, it tells us very little about the *multiplicative* structure of F. We shall now proceed to investigate the multiplicative structure. As we shall see, the central fact is that the multiplicative group of F is cyclic of order $q-1$.

Let $\alpha \in F$ be an arbitrary nonzero element of F, and consider the sequence of powers $1, \alpha, \alpha^2, \ldots, \alpha^n, \ldots$ of α. Each power α^i again lies in F, but since F contains only finitely many elements, the sequence must repeat. Let $\alpha^k = \alpha^{k+t}$ be the first repeat in the sequence. Then clearly $k = 0$; otherwise $\alpha^{k-1} = \alpha^{k+t-1}$ would be an earlier repeat. Thus $(1, \alpha, \ldots, \alpha^{t-1})$ are all distinct, but $\alpha^t = 1$. The integer $t \geq 1$ is called the *order* of α. This number will in general be different for different values of α; and given an element α, it may be difficult to calculate t. However, it turns out that we can say exactly *how many* elements of each order $t \geq 1$ are contained in F. Our first step in this direction is a special case of a famous theorem of Lagrange.

Theorem 5.2. *If t is the order of α, then t divides $q-1$.*

Proof: Let $F^\star$ denote the set of nonzero elements of F. Then $F^\star$ is a multiplicative group with $q-1$ elements, and $\{1, \alpha, \alpha^2, \ldots, \alpha^{t-1}\}$ is a subgroup with t elements. Lagrange's theorem tells us that the number of elements in a subgroup is always a divisor of the number of elements in the group; therefore t divides $q-1$, as promised. ∎

We pause in our development for an easy but vitally important lemma about solutions to polynomial equations over fields.

Lemma 5.3. *If $p(x)$ is a polynomial of degree m with coefficients in a field F, then the equation $p(x) = 0$ can have at most m distinct solutions in F.*

Proof: Induction on m; if $m = 1$ the equation is of the form $ax + b = 0$ which obviously has only the solution $x = -b/a$. If $m \geq 2$, and $p(x) = 0$ has no solution, QED. Otherwise let $p(\alpha) = 0$ and apply the division algorithm to divide $p(x)$ by $(x - \alpha)$: $p(x) = q(x)(x - \alpha) + r(x)$, with $\deg(r) < \deg(x - \alpha)$, or $r = 0$. This means r is an element of F. Then $p(\alpha) = 0$ implies $p(\alpha) = q(\alpha)(\alpha - \alpha) + r$, i.e., $r = 0$. Thus $p(x) = q(x)(x - \alpha)$ and any solution $\beta \neq \alpha$ to $p(x) = 0$ must be a solution to $q(x) = 0$; but $q(x)$ has degree $m - 1$ and so by induction there are at most $m - 1$ solutions to $q(x) = 0$, making at most m solutions to $p(x) = 0$. ∎

Example 5.1. Let Z_8 denote the ring of integers (mod 8), i.e. the set $\{0, 1, 2, \ldots, 7\}$ equipped with mod 8 addition and multiplication. In this ring the polynomial equation

$$x^2 - 1 = 0$$

has 4 distinct solutions, viz., $x = 1, 3, 5, 7$. This is a quadratic equation with four roots! This doesn't contradict Lemma 5.3, since Z_8 isn't a field. Still, this shows that Lemma 5.3 isn't true without some restriction. ∎

Notation. *From now on, we denote the order of $\alpha \in F$ by* $\mathrm{ord}(\alpha)$.

One consequence of Lemma 5.3 is this: If $\mathrm{ord}(\alpha) = t$, then every $\beta \in F$ such that $\beta^t = 1$ must be a power of α. This is because each of the t elements $\{1, \alpha, \ldots, \alpha^{t-1}\}$ satisfies the equation $x^t - 1 = 0$, and by Lemma 5.3 there can be no further solutions! But not every power of α has order t, as the following lemma shows.

Lemma 5.4. *If* $\mathrm{ord}(\alpha) = t$, *then* $\mathrm{ord}(\alpha^i) = t/\gcd(i, t)$.

Proof: We shall use the fact that for any $\beta \neq 0$

$$(5.4) \qquad \beta^s = 1 \quad \text{if and only if} \quad \mathrm{ord}(\beta) \mid s.$$

Let $d = \gcd(i, t)$. Then $\alpha^{i(t/d)} = \alpha^{t(i/d)} = (\alpha^t)^{(i/d)} = 1$. Thus by (5.4) $\mathrm{ord}(\alpha^i) \mid (t/d)$. Now suppose $s = \mathrm{ord}(\alpha^i)$. Then $\alpha^{is} = 1$ and so by (5.4) $t \mid is$. Since $d = \gcd(i, t)$, $ia + tb = d$ for certain integers a and b. Multiplying this equation by s we get $isa + tsb = ds$. But since $t \mid is$, it follows that $t \mid ds$, i.e., $(t/d) \mid s$; i.e., $(t/d) \mid \mathrm{ord}(\alpha^i)$. We have thus shown both $\mathrm{ord}(\alpha^i) \mid (t/d)$ and $(t/d) \mid \mathrm{ord}(\alpha^i)$. Hence $\mathrm{ord}(\alpha^i) = t/d$, as asserted. ∎

Example 5.2. Suppose $\mathrm{ord}(\alpha) = 12$, α being an element of some field F. We can compute the orders of $\alpha^i, i = 0, 1, \ldots, 11$, using Lemma 5.4; the work is summarized in the following table:

i	$\gcd(i, 12)$	$\mathrm{ord}(\alpha^i)$
0	12	1
1	1	12
2	2	6
3	3	4
4	4	3
5	1	12
6	6	2
7	1	12
8	4	3
9	3	4
10	2	6
11	1	12

This is a bit surprising. Given only the fact that there exists *at least one* element of order 12 in F, it follows that there are *exactly* 4 elements of order 12! Furthermore, there are exactly 2 elements of order 6, 2 of order 4, 2 of order 3, 4 of order 2, and 1 of order 1. Notice that to draw these conclusions, we only had do calculations with integers, and didn't need to know anything about F. It is remarkable that such sharp numerical results follow from the abstract axioms defining fields. ∎

In order to generalize the result of Example 5.2, we introduce the symbol $\phi(t)$ to denote the number of integers in the set $\{0, 1, \ldots, t-1\}$ which are relatively prime to t. This is Euler's famous "ϕ function." Note that since $\gcd(t, 1) = 1$ for all $t \geq 1$, then $\phi(t)$ is always at least 1. The exact value of

$\phi(t)$ is somewhat unpredictable, but for future reference we note that if t is prime, then $\phi(t) = t - 1$, since then every element in the set $\{0, 1, \ldots, t-1\}$ except 0 is relatively prime to t.

Theorem 5.5. *Let t be an integer, F a field. In F there are either no elements of order t, or exactly $\phi(t)$ elements of order t.*

Proof: If there are no elements of order t, there is nothing to prove. If $\text{ord}(\alpha) = t$, then as we observed above, every element of order t is in the set $\{1, \alpha, \ldots, \alpha^{t-1}\}$. But by Lemma 5.4, α^i will have order t if and only if $\gcd(i, t) = 1$. The number of such i is by definition $\phi(t)$. ∎

Combining Theorems 5.2 and 5.5, we see that if F is a finite field with q elements, and t is a positive integer, if t does not divide $q - 1$, there are no elements of order t; but if t does divide $q - 1$, there are either no elements of order t, or $\phi(t)$ elements of order t. We shall now prove that if t does divide $q-1$, there are always $\phi(t)$ elements of order t. Before giving a general proof, we consider another example.

Example 5.3. Suppose $q = 16$. Then $q - 1 = 15$ and by Theorem 5.2 the only possible values of t are $t = 1, 3, 5, 15$. For each of these values of t, the number of elements of order t is by Theorem 5.5 either 0 or $\phi(t)$. It is simple to compute $\phi(t)$ in each case, and we have:

t	$\phi(t)$
1	1
3	2
5	4
15	8

Now we notice a remarkable thing: the sum of the numbers in the $\phi(t)$ column equals 15, which is the total number of nonzero elements in F. It follows that in every case the number of elements of order t is $\phi(t)$: otherwise not all of the nonzero elements would be accounted for. This remarkable occurrence is no accident, as the next theorem shows. ∎

Theorem 5.6. *If n is any positive integer, then*

$$\sum_{d|n} \phi(d) = n, \tag{5.5}$$

where the notation in (5.5) indicates that the summation is to be extended over all positive divisors of n.

Proof: Let S_n denote the following set of rational fractions:

$$S_n = \{\tfrac{1}{n}, \tfrac{2}{n}, \cdots, \tfrac{n}{n}\},$$

and let T_n denote the subset of S_n consisting of the "reduced" fractions, i.e., those whose numerator is relatively prime to n. Then clearly $|S_n| = n$, $|T_n| = \phi(n)$ for all $n = 1, 2, 3, \ldots$. For example,

$$S_6 = \{\tfrac{1}{6}, \tfrac{2}{6}, \tfrac{3}{6}, \tfrac{4}{6}, \tfrac{5}{6}, \tfrac{6}{6}\},$$
$$T_6 = \{\tfrac{1}{6}, \tfrac{5}{6}\}.$$

Let us now imagine that we start with the set S_n, and reduce each fraction to lowest terms. Clearly each fraction $k/n \in S_n$ will, when reduced, have a unique numerator and denominator. Its denominator will be a divisor d of n, and its numerator e will be an integer $1 \leq e \leq d$ which is relatively prime to d. Again using $n = 6$ to illustrate,

$$\begin{array}{ll} \tfrac{1}{6} \to \tfrac{1}{6} & d = 6,\ e = 1 \\ \tfrac{2}{6} \to \tfrac{1}{3} & d = 3,\ e = 1 \\ \tfrac{3}{6} \to \tfrac{1}{2} & d = 2,\ e = 1 \\ \tfrac{4}{6} \to \tfrac{2}{3} & d = 3,\ e = 2 \\ \tfrac{5}{6} \to \tfrac{5}{6} & d = 6,\ e = 5 \\ \tfrac{6}{6} \to \tfrac{1}{1} & d = 1,\ e = 1. \end{array}$$

Conversely, if d is any positive divisor of n and if $1 \leq e \leq d$, with $\gcd(e, d) = 1$, the fraction e/d will appear as the reduced form of some fraction in S_n. As a result, S_n will decompose into the disjoint union of the sets T_d for all divisors d of n. Thus

$$S_n = \bigcup_{d|n} T_d,$$

and so

$$|S_n| = \sum_{d|n} |T_d|.$$

But since $|S_n| = n$, and $|T_d| = \phi(d)$, the proof is complete. ∎

Example 5.4. Theorem 5.6 can be used to aid in the calculation of $\phi(n)$. For example, consider $\phi(15)$. To compute $\phi(15)$, we could test each of the numbers $1, 2, \ldots, 15$ for relative-primeness with 15. Alternatively, however, Theorem 5.6 says that

$$\phi(1) + \phi(3) + \phi(5) + \phi(15) = 15.$$

But plainly $\phi(1) = 1$, $\phi(3) = 2$, and $\phi(5) = 4$. Thus $\phi(15) = 15 - 1 - 2 - 4 = 8$. What Theorem 5.6 gives us, in fact, is a recursive procedure for calculating $\phi(n)$. We will exploit this fully in Chapter 6 when we consider Möbius inversion. ∎

Combining Theorems 5.2, 5.5, and 5.6, we have the following important result.

Theorem 5.7. *Let F be a finite field with q elements, and let t be a positive integer. If $t \nmid (q-1)$, there are no elements of order t in F. If $t \mid (q-1)$ there are exactly $\phi(t)$ elements of order t in F.*

Proof: The first statement follows immediately from Theorem 5.2. Now for each positive divisor t of $q-1$, denote by $\psi(t)$ the number of elements of order t in F. Then since every element must have *some* order dividing $q-1$,

$$\sum_{t|q-1} \psi(t) = q - 1. \tag{5.6}$$

Combining (5.6) and (5.5) we see that

$$\sum_{t|q-1} (\phi(t) - \psi(t)) = 0.$$

But by Theorem 5.5, $\phi(t) - \psi(t) \geq 0$, for all t. Thus $\phi(t) = \psi(t)$ for all $t \mid q-1$. ∎

Corollary. *In every finite field, there exists at least one element (in fact, exactly $\phi(q-1)$ elements) of order $q-1$. Hence, the multiplicative group of any finite field is cyclic.*

Definition. *An element of multiplicative order $q-1$, i.e., a generator of the cyclic group $F^\star = F - \{0\}$, is called a* **primitive root** *of the field F.*

Example 5.5. Consider the field $F_7 = Z \bmod 7$, whose elements we take to be $\{0, 1, 2, 3, 4, 5, 6\}$. The powers of 2 are $2^0 = 1$, $2^1 = 2$, $2^2 = 4$, $2^3 = 1$. Thus neither 2 nor any of its powers is a primitive root in F_7. Since $\{1, 2, 4\}$ are not primitive roots, we try 3. The powers of 3 are summarized in the following table:

i	3^i	$\text{ord}(3^i)$
0	1	1
1	3	6
2	2	3
3	6	2
4	4	3
5	5	6
6	1	(repeats)

Thus $\text{ord}(3) = 6$, i.e., 3 is a primitive root in F_7. Now using Lemma 5.4 we can compute the order of each nonzero element in F_7, in terms of its "base 3 logarithm," or as it is usually called in this context, its *index*. We see that there are indeed $\phi(1) = 1$ elements of order 1, $\phi(2) = 1$ of order 2, $\phi(3) = 2$ of order 3, and $\phi(6) = 2$ of order 6. ∎

In the preceding example, we found a primitive root by trial and error. In a larger finite field it is desirable to have a systematic procedure. We now present an algorithm for finding primitive roots in an arbitrary finite field. It is due to Gauss himself.

Gauss's algorithm produces a sequence of field elements $\alpha_1, \alpha_2, \ldots, \alpha_k$ with $\text{ord}(\alpha_1) < \text{ord}(\alpha_2) < \cdots < \text{ord}(\alpha_k) = q-1$. In fact it will turn out that $\text{ord}(\alpha_i) \mid \text{ord}(\alpha_{i+1})$, for $i = 1, 2, \ldots, k-1$.

Gauss's algorithm:

G1. Set $i = 1$, and let α_1 be an arbitrary nonzero element of F. Let $\text{ord}(\alpha_1) = t_1$.

G2. If $t_i = q-1$, stop: α_i is the desired primitive root.

G3. Otherwise choose a nonzero element $\beta \in F$ which is not a power of α_i. Let $\text{ord}(\beta) = s$. If $s = q-1$ set $\alpha_{i+1} = \beta$ and stop.

G4. Otherwise find $d \mid t_i$ and $e \mid s$ with $\gcd(d, e) = 1$ and $d \cdot e = \text{lcm}(t_i, s)$. Let $\alpha_{i+1} = \alpha_i^{t_i/d} \cdot \beta^{s/e}$, $t_{i+1} = \text{lcm}(t_i, s)$. Increase i by 1 and go to $G2$.

Notes on the algorithm:

- In step G3, observe that the order s of β will not be a divisor of t_i, since all solutions to $x^{t_i} = 1$ must be powers of α_i. Hence $\text{lcm}(t_i, s)$ will be a multiple of t_i, strictly greater than t_i.
- In step G4 it is essentially asserted that given two integers m and n it is possible to find $d \mid m$, $e \mid n$ with $\gcd(d, e) = 1$, and $de = \text{lcm}(m, n)$. For example, with $m = 12$, $n = 18$, the proper choice is $d = 4$, $e = 9$. We leave as an exercise the demonstration that this decomposition is always possible.
- In step G4, the element $\alpha_i^{t_i/d}$ will have order d, and the element $\beta^{(s/e)}$ will have order e (see Lemma (5.4)). It follows that the product of these two elements will have order $d \cdot e = \text{lcm}(t_i, s)$, because of the following lemma.

Lemma 5.8. *Let* $\text{ord}(\alpha) = m$, $\text{ord}(\beta) = n$, *with* $\gcd(m, n) = 1$. *Then* $\text{ord}(\alpha\beta) = mn$.

Proof: (We use Eq. (5.3)). Suppose $t = \text{ord}(\alpha\beta)$. Then $1 = (\alpha\beta)^t = (\alpha\beta)^{tn} = \alpha^{tn}\beta^{tn} = \alpha^{tn}$, since $\text{ord}(\beta) = n$. Thus $\alpha^{tn} = 1$, i.e., $m \mid tn$. But since $\gcd(m, n) = 1$, we have $m \mid t$. Similarly $n \mid t$. Again, since $\gcd(m, n) = 1$, we have $mn \mid t$, i.e., $mn \mid \text{ord}(\alpha\beta)$. Conversely $(\alpha\beta)^{mn} = \alpha^{mn}\beta^{mn} = 1$, so $\text{ord}(\alpha\beta) \mid mn$. ∎

Thus we see that the new element α_{i+1} generated in step G4 will have an order t_{i+1} which is a proper multiple of $\text{ord}(\alpha_i)$, and so the process must eventually terminate with a primitive root.

In Example 5.5 above, we used (implicitly) Gauss's algorithm and discovered $\alpha_1 = 2$, $t_1 = 3$, $\alpha_2 = 3$, $t_2 = 6$. We never got to G4. The following example will test Gauss's algorithm more strenuously.

Example 5.6. We construct a field F with 25 elements using the polynomial $x^2 - 2$, which is easily seen to be irreducible in $F_5[x]$ (there is no square root of 2 in F_5). The elements of F we view as two-dimensional vectors (a, b) with $a, b \in F_5 = \{0, 1, 2, 3, 4\}$. Addition in F is componentwise:

$$(a_1, b_1) + (a_2, b_2) = (a_1 + a_2, b_1 + b_2).$$

To define multiplication we view (a, b) as the linear polynomial $ax + b$, and use the rule

$$\begin{aligned}(a_1x + b_1) \cdot (a_2x + b_2) &= (a_1a_2x^2 + (a_1b_2 + a_2b_1)x + b_1b_2) \\ &= (a_1b_2 + a_2b_1)x + (b_1b_2 + 2a_1a_2),\end{aligned}$$

which follows from the ordinary rules for multiplying polynomials and the fact that $x^2 \equiv 2 \pmod{x^2 - 2}$. Thus in the field F, the rule for multiplying (a_1, b_1) and (a_2, b_2) is

$$(a_1, b_1) \cdot (a_2, b_2) = (a_1b_2 + a_2b_1, 2a_1a_2 + b_1b_2)$$

Let us use Gauss's algorithm to find a primitive root in this field. We begin by taking the element $(1, 0)$ as α_1. To compute $\text{ord}(\alpha_1)$, we compute a little table of its powers:

i	α_1^i
0	$(0,1)$
1	$(1,0)$
2	$(0,2)$
3	$(2,0)$
4	$(0,4)$
5	$(4,0)$
6	$(0,3)$
7	$(3,0)$
8	$(0,1)$

Thus $\operatorname{ord}(\alpha_1) = 8$, i.e., in Gauss's algorithm we have $t_1 = 8$. Since $t_1 \neq q-1 = 24$, α_1 is not a primitive root and we proceed to step G3. We must choose an element β which is not a power of α_1, say $\beta = (1,1)$. Then we have

i	β^i
0	$(0,1)$
1	$(1,1)$
2	$(2,3)$
3	$(0,2)$
4	$(2,2)$
5	$(4,1)$
6	$(0,4)$
7	$(4,4)$
8	$(3,2)$
9	$(0,3)$
10	$(3,3)$
11	$(1,4)$
12	$(0,1)$.

Thus $\operatorname{ord}(\beta) = 12$, and so

$$\begin{aligned} \alpha_1 &= (1,0), \qquad t_1 = 8 \\ \beta &= (1,1), \qquad s = 12, \end{aligned}$$

and we enter step G4. Here we take $d = 8$, $e = 3$, since $8 \cdot 3 = \operatorname{lcm}(8,12) = 24$.

Therefore $\alpha_2 = \alpha_1 \cdot \beta^4 = \alpha_1 \cdot (2\alpha_1 + 2) = 2\alpha_1^2 + 2\alpha_1 = 2\alpha_1 + 4$ has order 24. Thus $t_2 = 24$, we return to G2, and stop; $2\alpha_1 + 4$ is a primitive root in F. ∎

So much, temporarily, for primitive roots. We come now to a new topic, *minimal polynomials.*

Let F be a finite field; by Theorem 5.1 F has p^m elements for some prime p and some positive integer m. Moreover, in the proof of Theorem 5.1 we saw that F could be viewed as an m–dimensional vector space over F_p. Now let α be an arbitrary element of F. Consider the $m+1$ elements

$$1, \alpha, \alpha^2, \ldots, \alpha^m$$

of F. Since F has dimension m over F_p, it follows that these $m+1$ elements must be *linearly dependent* over F_p, i.e., there exist $m+1$ elements of F_p, say $A_0, A_1, \ldots, A_m$, not all zero, such that

$$A_0 + A_1\alpha + \cdots + A_m\alpha^m = 0.$$

In other words, if $A(x) = A_0 + A_1x + \cdots + A_mx^m$, α satisfies the polynomial equation

$$A(x) = 0. \tag{5.7}$$

Of course, α may satisfy other polynomial equations, and so we define $S(\alpha)$ to be the set of all such polynomials:

$$S(\alpha) = \{f(x) \in F_p(x) : f(\alpha) = 0\}.$$

We have seen that $S(\alpha)$ is not empty, and indeed that $S(\alpha)$ contains at least one polynomial of degree $\leq m$.

Let $p(x)$ be a nonzero polynomial of least degree in $S(\alpha)$; and let $f(x)$ be any polynomial in $S(\alpha)$. By the division algorithm for polynomials, there exist polynomials $q(x)$, $r(x)$ such that

$$f(x) = q(x)p(x) + r(x), \qquad \deg(r) < \deg(p).$$

But $f(\alpha) = p(\alpha) = 0$, and so $r(\alpha) = 0$ as well. This contradicts the fact that $\deg(p)$ is minimal unless $r(x) \equiv 0$, and so we conclude,

$$p(x) \mid f(x) \qquad \text{for all } f(x) \in S(\alpha). \tag{5.8}$$

The polynomial $p(x)$ in (5.8) we call a *minimal polynomial* of α with respect to the field F. If we stipulate that $p(x)$ must be monic, then (5.8) shows that $p(x)$ is unique.

We observe finally that the monic polynomial $p(x)$ of α must be *irreducible*. This is because if, say, $p(x) = a(x) \cdot b(x)$, then on account of $p(\alpha) = 0$, we would necessarily have $a(\alpha) = 0$ or $b(\alpha) = 0$, which would contradict the minimality of the degree of $p(x)$.

This discussion has led us to the following important result.

Theorem 5.9. *Suppose F is a field with p^m elements. Associated with each $\alpha \in F$, there is a unique monic polynomial $p(x) \in F_p(x)$, with the following properties:*

(a) $p(\alpha) = 0$,
(b) $\deg(p) \le m$,
(c) If $f(x)$ is another polynomial in $F_p(x)$ with $f(\alpha) = 0$, then $p(x) \mid f(x)$. ∎

The polynomial described in Theorem 5.9 is called *the minimal polynomial of* α with respect to the subfield F_p of F.

Example 5.7. We return to the field of order 25 in Example 5.6, and compute the minimal polynomials for some of the elements. We begin by considering the powers of the element $(1, 0)$, which we shall from now on denote simply by α.

Consider first $\alpha^0 = 1$. Here there is no problem; the minimal polynomial is $x - 1$. More generally, if $\alpha \in F$ is in fact an element of F_p, its minimal polynomial will always be $x - \alpha$.

Next consider α itself. Now $\alpha \notin F_5$, so $x - \alpha$ will not do (its coefficients aren't all in F_p). However, the powers 1, α, α^2, viewed as three vectors in the two-dimensional vector space F, must be linearly dependent, and a linear dependence among them will immediately give us a quadratic equation satisfied by α:

$$\begin{aligned} 1 &= (0,1) \\ \alpha &= (1,0) \\ \alpha^2 &= (0,2). \end{aligned}$$

Clearly $\alpha^2 - 2 \cdot 1 = 0$, and so the minimal polynomial for α is $x^2 - 2$, which, not by accident, is the defining polynomial for the field. Continuing in this way, we get the following table of minimal polynomials.

i	α^i	minimal polynomial
0	$(0,1)$	$x - 1$
1	$(1,0)$	$x^2 - 2$
2	$(0,2)$	$x - 2$
3	$(2,0)$	$x^2 - 3$
4	$(0,4)$	$x - 4$
5	$(4,0)$	$x^2 - 2$
6	$(0,3)$	$x - 3$
7	$(3,0)$	$x^2 - 3$

Now let's compute the minimal polynomial of the primitive root $2\alpha + 4 = \beta$ that we found in Example 5.6. We begin by computing its first three powers:

$$\begin{aligned} 1 &= (0,1) \\ \beta &= (2,4) \\ \beta^2 &= (1,4). \end{aligned}$$

Hence $\beta^2 = 3 \cdot \beta + 2 \cdot 1$, and so the minimal polynomial of β is $x^2 - 3x - 2$ or $x^2 + 2x + 3$. This polynomial is by Theorem 5.9 irreducible. If we had used it, rather than $x^2 - 2$, to build our field, then the element $\alpha = (1,0)$ would have turned out to be a *primitive root.* This would have been a much more convenient representation of the field. In general, the minimal polynomial of a primitive root in F is called a *primitive polynomial.* We will have much more to say about primitive polynomials later on, for example in Chapter 10, when we study m-sequences. ∎

In the remainder of this chapter we plan to give a systematic account of minimal polynomials in finite fields, an account which will perhaps clarify some of the mysteries of Example 5.7.

Thus let F be a finite field, and let k be a subfield of F, i.e., a subset of F which is itself a field. (For example, k could be the prime subfield F_p; but it could also be a "bigger" subfield.)

Denote the number of elements in k by q. The number q is, by Theorem 5.1, a power of a prime p; say $q = p^m$. Just as in the proof of Theorem 5.1, we can view F as a vector space over k and conclude that the number of elements in F is a power of the number of elements in k. Thus $|F| = q^n = p^{n \cdot m}$ for some integer n.

Now let $\alpha \in F$ be arbitrary. The minimal polynomial of α *with respect to* k is the unique monic polynomial $p(x)$ with the properties cited in Theorem 5.9, except that we replace the prime field F_p with k. We propose to get a more-or-less explicit expression for $p(x)$. But first we need several lemmas.

Lemma 5.10. *An element $\beta \in F$ lies in the subfield k if and only if $\beta^q = \beta$. In particular, every element of k satisfies the equation $x^q - x = 0$.*

Proof: Let $\beta \in k$. If $\beta = 0$, clearly $\beta^q = \beta$. Otherwise let $t = \operatorname{ord}(\beta)$. Then by Theorem 5.2, $t \mid q-1$ and so (see Eq. (5.3)) $\beta^{q-1} = 1$. Thus $\beta^q = \beta$. Hence the q elements of k provide q distinct solutions to $x^q - x = 0$. By Lemma 5.3, there can be no other solutions. ∎

Lemma 5.11. *If p is prime, the binomial coefficient $\binom{p}{k}$ is divisible by p for $1 \le k \le p-1$.*

Proof: By definition,

$$\binom{p}{k} = \frac{p(p-1)\cdots(p-k+1)}{k(k-1)\cdots 1}.$$

The numerator of the fraction is divisible by p, but the denominator is not. ∎

Lemma 5.12. *Let $\alpha_1, \alpha_2, \ldots, \alpha_t$ be elements in any field (finite or not) of characteristic p. Then*

$$(\alpha_1 + \alpha_2 + \cdots + \alpha_t)^{p^k} = \alpha_1^{p^k} + \alpha_2^{p^k} + \cdots + \alpha_t^{p^k}$$

for $k = 1, 2, 3, \ldots$.

Proof: We prove the case $t = 2$; the general case follows easily by induction. For $t = 2$ the claim is that

$$(\alpha + \beta)^{p^k} = \alpha^{p^k} + \beta^{p^k}. \tag{5.9}$$

Consider first $k = 1$. By the binomial theorem

$$(\alpha + \beta)^p = \alpha^p + \binom{p}{1}\alpha^{p-1}\beta + \cdots + \binom{p}{p-1}\alpha\beta^{p-1} + \beta^p.$$

By Lemma 5.11, the binomial coefficients are all integers divisible by p; thus in a field of characteristic p they are zero. Hence

$$(\alpha + \beta)^p = \alpha^p + \beta^p. \tag{5.10}$$

This is the case $k = 1$ of Eq. (5.9). Having proved (5.9) for k, raise (5.9) to the pth power and use (5.10). The result is (5.9) with k replaced by $k + 1$. Hence (5.9) is true by induction. ∎

Armed with the preceding lemmas, we can attack the problem of minimal polynomials. Our first step is to notice that if $p(x) = p_0 + p_1 x + p_2 x^2 + \cdots + p_d x^d \in k(x)$ is such that $p(\alpha) = 0$, i.e., if

$$\sum_{k=0}^{d} p_k \alpha^k = 0, \qquad p_k \in k, \tag{5.11}$$

then also $p(\alpha^q) = 0$. This is because if we raise (5.11) to the qth power, we get

$$\begin{aligned} 0 &= \left(\sum_{k=0}^{d} p_k \alpha^k\right)^q \\ &= \sum_{k=0}^{d} p_k^q \alpha^{qk} && \text{(by Lemma 5.12)} \\ &= \sum_{k=0}^{d} p_k (\alpha^q)^k && \text{(by Lemma 5.10)} \\ &= p(\alpha^q). \end{aligned}$$

Thus if α is a zero of $p(x)$, so is α^q. Repeating this argument with α^q in place of α, we find that α^{q^2} is also a zero of $p(x)$, and so forth. The elements

$$\alpha, \alpha^q, \alpha^{q^2}, \alpha^{q^3}, \ldots \tag{5.12}$$

are called the *conjugates* of α (with respect to the sub-field k). Of course, there are only finitely many elements in the list (5.12), and so there must be repeats. Assume $\alpha \neq 0$, and let

$$\alpha^{q^d} = \alpha^{q^j}, \qquad \text{where } j < d \tag{5.13}$$

be the first repeat in the sequence (5.12). Then

$$\begin{aligned} 1 &= \alpha^{q^d - q^j} \\ &= \alpha^{q^j (q^{d-j} - 1)}, \end{aligned}$$

and so by (5.3) $\operatorname{ord}(\alpha) \mid q^j(q^{d-j} - 1)$. But $\operatorname{ord}(\alpha) \mid q^n - 1$ by Theorem 5.2, which implies $\gcd(\operatorname{ord}(\alpha), q^j) = 1$. Hence $\operatorname{ord}(\alpha) \mid q^{d-j} - 1$. But this implies $\alpha^{q^{d-j}} = \alpha$, and since (5.13) is assumed to be the *first* repeat, necessarily $j = 0$.

We have just concluded that $\alpha^{q^d} = \alpha$. But since F has q^n elements, we know from Lemma 5.10 that $\alpha^{q^n} = \alpha$. Thus by Corollary 2.4, $\alpha^{\gcd(d,n)} = \alpha$. This contradicts the fact that α^{q^d} is the earliest repeat, unless $\gcd(d,n) = d$, i.e., unless d is a divisor of n. In summary, *the number of distinct conjugates of α is a divisor of n.*

The number d of conjugates of α is called the *degree* of α; we have seen above that d is the smallest positive integer such that

$$q^d \equiv 1 \pmod{t}, \tag{5.14}$$

where $t = \text{ord}(\alpha)$. Summarizing, we have proved the following.

Theorem 5.13. *The number d of conjugates of α is a divisor of n. This number d can be determined as the smallest integer such that (5.14) holds. Furthermore, if t is arbitrary and if $t = \alpha \cdot d + r$, $0 \le r \le d-1$, then*

$$\alpha^{q^t} = \alpha^{q^r}.$$

We now know that the minimal polynomial for α must have at least d zeros, viz., $\alpha, \alpha^q, \alpha^{q^2}, \ldots, \alpha^{q^{d-1}}$. Thus if we define

$$f_\alpha(x) = (x-\alpha)(x-\alpha^q)\cdots(x-\alpha^{q^{d-1}}), \tag{5.15}$$

it follows that $f_\alpha(x)$ must be a divisor of α's minimal polynomial. We shall now show that $f_\alpha(x)$ is in fact *equal to* the minimal polynomial of α. To do this, we expand $f_\alpha(x)$ in powers of x:

$$\begin{aligned}(x-\alpha)(x-\alpha^q)\cdots(x-\alpha^{q^{d-1}}) = A_d x^d + A_{d-1}x^{d-1} + \cdots \\ + A_1 x + A_0\end{aligned} \tag{5.16}$$

where in (5.16) each coefficient A_i is an element of F. We now raise both sides of (5.16) to the qth power:

$$\begin{aligned}(x-\alpha)^q(x-\alpha^q)^q\cdots(x-\alpha^{q^{d-1}})^q = A_d^q x^{qd} + A_{d-1}^q x^{q(d-1)} + \cdots \\ + A_1^q x^q + A_0^q,\end{aligned} \tag{5.17}$$

where to get the right side of (5.17) we have used Lemma 5.12. Now again by Lemma 5.12, $(x-\beta)^q = x^q + (-1)^q\beta^q = x^q - \beta^q$. (Note separate arguments needed for even and odd q.) Thus the left side of (5.17) is equal to

$$(x^q - \alpha^q)(x^q - \alpha^{q^2})\cdots(x^q - \alpha^{q^{d-1}})(x^q - \alpha) = f_\alpha(x^q). \tag{5.18}$$

But from the expansion $f_\alpha(x) = \sum A_i x^i$ we have

$$f_\alpha(x^q) = A_d x^{qd} + A_{d-1}x^{q(d-1)} + \cdots + A_1 x^q + A_0. \tag{5.19}$$

But since the polynomials on the right side of (5.17) and (5.19) are equal, we must have $A_i^q = A_i$, $i = 0, 1, \ldots, d-1$, i.e., (Lemma 5.10) $A_i \in k$ for all i. To summarize: we have found that the polynomial $f_\alpha(x)$, as defined by (5.15), is a divisor of the minimal polynomial of α, and that the coefficients of $f_\alpha(x)$ all lie in the small field k. By Theorem 5.9, we conclude that $f_\alpha(x)$ must be the minimal polynomial of α. We state this important conclusion in a theorem.

Theorem 5.14. *Let F be a field with q^n elements, and k be a q-element subfield. If $\alpha \in F$, then the minimal polynomial of α with respect to the subfield k is*

$$f_\alpha(x) = (x-\alpha)(x-\alpha^q)\cdots(x-\alpha^{q^{d-1}}),$$

where $d =$ the degree of α with respect to k, as determined by (5.14). ∎

Example 5.8. Consider the case $q = 2, n = 4$. It is known that $x^4 + x + 1$ is irreducible and primitive over $F_2(x)$, and we shall construct a field with 16 elements using it. If we denote by α the field element corresponding to $x \bmod x^4 + x + 1$, i.e., to the vector $(0,0,1,0)$, we calculate:

i	α^i	ord(α^i)	deg(α)	minimal polynomial
0	0001	1	1	$x+1$
1	0010	15	4	$(x-\alpha)(x-\alpha^2)(x-\alpha^4)(x-\alpha^8)$ $=x^4+x+1$
2	0100	15	4	x^4+x+1
3	1000	5	4	$(x-\alpha^3)(x-\alpha^6)(x-\alpha^{12})(x-\alpha^9)$ $=x^4+x^3+x^2+x+1$
4	0011	15	4	x^4+x+1
5	0110	3	2	$(x-\alpha^5)(x-\alpha^{10})=x^2+x+1$
6	1100	5	4	$x^4+x^3+x^2+x+1$
7	1011			
8	0101			
9	1010			
10	0111			(rest of table left as exercise)
11	1110			
12	1111			
13	1101			
14	1001			
15	0001			

Notes on these calculations:

• $i = 0$: The minimal polynomial of 1 in any field is $x - 1$; of course in characteristic 2, $x - 1$ and $x + 1$ are the same.

• $i = 1$: The minimal polynomial of α will always turn out to be the polynomial used to define the field. In this example, we used $x^4 + x + 1$ to define the field, which forces α to satisfy the equation $\alpha^4 = \alpha + 1$. The skeptical reader can multiply out $(x-\alpha)(x-\alpha^2)(x-\alpha^4)(x-\alpha^8)$ the long way if he/she doubts this!

• $i = 2$: Since $\alpha, \alpha^2, \alpha^4, \alpha^8$ are all conjugates, they all have the same minimal polynomial. Since we already know that $x^4 + x + 1$ is the minimal polynomial for α, it is also the minimal polynomial for α^2.

• $i = 3$: One could multiply $(x-\alpha^3)(x-\alpha^6)(x-\alpha^{12})(x-\alpha^9)$ out by brute force to verify that the minimal polynomial of α^3 is $x^4+x^3+x^2+x+1$, but an easier and more systematic approach is to seek a linear dependence among the first 5 powers of α^3. Denoting α^3 temporarily by β, we have

$$\begin{aligned} 1 &= 0001 \\ \beta &= 1000 \\ \beta^2 &= 1100 \\ \beta^3 &= 1010 \\ \beta^4 &= 1111 \end{aligned}$$

We want to find a linear dependence among these five four-dimensional vectors. (Of course in this particular example it's not hard to find the linear dependence "by eye," but let's adopt a systematic approach that will work in general.) Such a linear dependence is just a vector in the nullspace of the column space of the above 5×4 matrix, and we can find it by performing elementary column operations. By adding the second column to the first, we get

$$\begin{matrix} 0 & 0 & 0 & 1 \\ 1 & 0 & 0 & 0 \\ 0 & 1 & 0 & 0 \\ 1 & 0 & 1 & 0 \\ 0 & 1 & 1 & 1. \end{matrix}$$

Now add the third column to the first:

$$\begin{matrix} 0 & 0 & 0 & 1 \\ 1 & 0 & 0 & 0 \\ 0 & 1 & 0 & 0 \\ 0 & 0 & 1 & 0 \\ 1 & 1 & 1 & 1 \end{matrix}$$

This matrix is in column-reduced form and it is now obvious that the sum of all the rows is zero, i.e., $1 + \beta + \beta^2 + \beta^3 + \beta^4 = 0$. Hence the minimal polynomial for $\beta = \alpha^3$ is $x^4 + x^3 + x^2 + x + 1$, as asserted. Alternatively, note that β is a fifth root of 1 but not equal to 1, so β must be a zero of

$$\frac{x^5 - 1}{x - 1} = x^4 + x^3 + x^2 + x + 1.$$

We leave the completion of the table as an exercise. ∎

Problems for Chapter 5.

1. Prove (5.2a) and (5.2b).

2. What is the smallest prime p such that the field F_p has both 2 and 3 as primitive roots?

3. Consider the field $F_{73} = Z \bmod 73$.

a. Calculate ord(2) and ord(3).

b. Find an element of order lcm(ord(2), ord(3)).

c. Let α be the number found in part (b). What is the smallest integer in $\{1, 2, \ldots, 72\}$ which is not a power of α? What is the order of this number?

d. Compute $\sqrt[30]{64}$ and $\sqrt[30]{59}$ in F_{73}. [Hint: Let α be a prim. root in F_{73}. Then $\beta^{30} = \gamma$ is equivalent to $30x \equiv y \pmod{72}$, if $\beta = \alpha^x$ and $\gamma = \alpha^y$.]

4. In Example 5.1 we saw that the equation $x^2 - 1 = 0$ has four roots in the ring Z_8, of arithmetic mod 8.

a. How many roots does $x^2 - 1 = 0$ have in the ring Z_{16}?

b. How many roots does $x^2 - 1 = 0$ have in the ring Z_{2^n}?

5. Let p be a prime number. What is $\phi(p^m)$? [Hint: Use Theorem 5.6.]

6. In Example 5.6 we asserted that the polynomial $x^2 - 2$ is irreducible over the field F_5. Verify this, and more generally determine which of the polynomials $x^2 - a$, where $a \in F_5$, is irreducible. For those polynomials which are not irreducible, give the factorization into irreducible factors.

7. Consider the field $GF(49)$, generated by $x^2 - 3 \pmod 7$. Let α denote the field element corresponding to the equivalence class of $x \pmod{x^2 - 3}$.

a. What is the order of α in this field?

b. Find a primitive root, and express α as a power of it.

c. Find the minimal polynomial of the primitive root found in part (b).

8. Build the field $GF(27)$, using the primitive polynomial $x^3 + 2x + 1$. List each of the 27 elements, and for each element, find its minimal polynomial and its order.

9. This problem relates to Step G4 in Gauss's algorithm.

a. Given two integers m and n show that it is always possible to find $d \mid m$, $e \mid n$ with $\gcd(d, e) = 1$, and $de = \operatorname{lcm}(m, n)$.

b. If $m = 150$ and $n = 225$, what is the the proper choice for d and e?

10. In Example 5.6 we found minimal polynomials for 9 of the 25 elements in the field constructed in Example 5.4, viz., the 8 powers of α and the primitive root $2\alpha + 4$. Find the minimal polynomials for the other 16 field elements.

11. In the text we only proved Lemma 5.12 for the case $t = 2$. Give the proof for general t, using induction.

12. In Example 5.8 we computed the minimal polynomials for the first seven powers of α, i.e., for $1, \alpha, \ldots, \alpha^6$. Compute the minimal polynomials for the remaining 8 powers of α, viz., $\alpha^7, \ldots, \alpha^{14}$.

13. In Lemmas 5.12 and 5.13, we used the binomial theorem, viz.

$$(a+b)^n = \sum_{k=0}^{n} \binom{n}{k} a^k b^{n-k}.$$

However, the binomial coefficients are integers, not elements in the finite field k. Explain how the binomial theorem can be modified so that it does hold in a finite field. [Hint: Use the sequence (u_n) defined in (5.1).]

14. When the fraction 1/7 is expressed as a repeating decimal, viz.

$$\tfrac{1}{7} = 0.14285714\cdots,$$

the period is 6, whereas the representation of 1/11, viz.

$$\tfrac{1}{11} = 0.090909\cdots,$$

repeats with period 2. Show that in general, the period of the decimal representation of the fraction k/p, where p is an odd prime (except 5) and $1 \leq k \leq p-1$, is equal to the order of 10 in the field F_p. Why is 5 exceptional?

Chapter 6

Finite Fields Exist and are Unique

Let us recapitulate our results so far. We know that there is, for every prime p, a finite field with p elements, viz. the integers (mod p). This field we have denoted by F_p. Furthermore, we know by Theorem 4.1 that provided there is an irreducible polynomial of degree m over F_p, there exists a field with p^m elements. On the other hand, by Theorem 5.1, there can be no finite field with a number of elements which is *not* a power of some prime number. This leaves two obvious questions:

A. For which values of p and m do irreducible polynomials exist?

B. Are there any other kinds of finite fields, i.e., ones that are not constructed by using Theorem 4.1?

In this chapter, we'll answer both questions completely. The answer to Question A is "For all values of p and m", and the answer to Question B is "No." These answers will follow from a study of how certain special polynomials *factor* over certain finite fields.

For our first result, we assume that k is a finite field with q elements. (For example q could be any prime number, or $q = 8$, as in Example 4.8, or $q = 25$, as in Example 5.4.) We consider now the polynomial $x^{q^n} - x$. By Theorem 3.1 we know that this polynomial will factor uniquely into a product of irreducible monic polynomials over k. The next theorem tells us something more about this factorization.

Theorem 6.1.

$$x^{q^n} - x = \prod_{d|n} V_d(x),$$

where $V_d(x)$ is the product of all monic irreducible polynomials in $k[x]$ of degree d.

Proof: Let d be a divisor of n and let $f(x)$ be a monic irreducible polynomial of degree d over k. Form the field $F = k[x] \pmod{f(x)}$; then F has q^d elements. Denote by α the element of F corresponding to the (mod f) equivalence class containing x; i.e., $\alpha = \overline{x}$. Then by e.g., Lemma 5.10, $\alpha^{q^d} = \alpha$, and this is equivalent to the statement

$$f(x) \mid (x^{q^d} - x). \tag{6.1}$$

Since by Corollary 2.4, $(x^{q^d} - x) \mid (x^{q^n} - x)$ iff $d \mid n$, it follows that every irreducible polynomial whose degree divides n divides $x^{q^n} - x$.

Now suppose $f(x) \mid (x^{q^n} - x)$. Then (6.1) also holds, if $d = \deg(f)$, and so $f(x) \mid \gcd(x^{q^n} - x, x^{q^d} - x) = x^{q^e} - x$ where $e = \gcd(d, n)$ by Corollary 2.4 again. In terms of the field $F = k[x] \pmod{f(x)}$, with $\alpha = \overline{x}$, this implies that

$$\alpha^{q^e} = \alpha. \tag{6.2}$$

Since every polynomial of degree $\leq d-1$ can be written as a linear combination of $1, x, x^2, \ldots, x^{d-1}$, it follows that every element of $F = k[x] \pmod{f(x)}$ can be written as a linear combination of $1, \alpha, \alpha^2, \ldots, \alpha^{d-1}$. Thus let $\beta = A_0 + A_1\alpha + \cdots + A_{d-1}\alpha^{d-1}$ be an arbitrary element of F. Then by Lemma 5.12,

$$\begin{aligned} \beta^{q^e} &= A_0^{q^e} + A_1^{q^e}\alpha^{q^e} + \cdots + A_{d-1}^{q^e}(\alpha^{q^e})^{d-1} \\ &= A_0 + A_1\alpha + \cdots + A_{d-1}\alpha^{d-1}, \text{ by (6.2)}. \end{aligned}$$

Thus not only α, but in fact *every* element $\beta \in F$ satisfies the equation $x^{q^e} - x = 0$. By Lemma 5.3 this equation can have at most q^e solutions, and so $e \geq d$. But $e = \gcd(d, n)$ is a divisor of d so $e = d$. This means that d is a divisor of n. It follows that every irreducible divisor of $x^{q^n} - x$ has degree dividing n.

The proof will be complete if we can show that $x^{q^n} - x$ has no repeated factors. To do this we will use the fact that a polynomial $P(x)$ has no repeated factors if and only if $\gcd(P(x), P'(x)) = 1$, where $P'(x)$ is the *formal derivative* of $P(x)$. The formal derivative of $P(x) = P_0 + P_1 x + \cdots + P_n x^n$ is defined to be $P_1 + 2P_2 x + \cdots + nP_n x^{n-1}$, i.e., just what you would get by applying the ordinary freshman calculus rules. But we will leave the investigation of formal derivatives as a problem, and use the result here. The formal derivative of $x^{q^n} - x$ is -1, since the integer q^n, i.e., $1 + 1 + \cdots + 1$ (q^n times) is equal to zero in a field with $q = p^m$ elements. Thus $x^{q^n} - x$ is relatively prime to its derivative and so can have no repeated factors.

We have therefore shown that $f(x) \mid (x^{q^n} - x)$ iff $\deg(f) \mid n$, and that $f(x)$ is squarefree. This completes the proof of Theorem 6.1. ∎

Example 6.1. Let $q = 2$, $n = 4$. Theorem 6.1 implies that $x^{16} + x$ is the product of all F_2—irreducible polynomials of degrees 1, 2, and 4. Indeed, by looking in tables or by using techniques to be developed in Chapter 7,

$$x^{16} + x = (x^4 + x + 1) \cdot (x^4 + x^3 + 1) \cdot (x^4 + x^3 + x^2 + x + 1) \cdot (x^2 + x + 1) \cdot (x + 1) \cdot x.$$

Multiplying these polynomials together, we find

$$V_1(x) = x^2 + x$$
$$V_2(x) = x^2 + x + 1$$
$$V_4(x) = x^{12} + x^9 + x^6 + x^3 + 1.$$

∎

By comparing the degrees of the two sides of Theorem 6.1, we get an extremely important result.

Corollary 6.2.

$$q^n = \sum_{d \mid n} d I_d,$$

where I_d denotes the number of distinct monic irreducible polynomials of degree d. ∎

The result of Corollary 6.2 is so important that we propose to give a second proof of it, a proof which is essentially due to Euler, and which when appropriately generalized becomes one of the most powerful tools of number theory, the Euler Product technique.

The basic idea is that we can, in principle, use the unique factorization theorem to calculate I_d, the number of irreducible monic polynomials of degree d. (The underlying field is understood to have q elements.) For example, if $q = 2$, $I_1 = 2$ since there are only two monic polynomials of degree 1 (x and $x + 1$), and they are both plainly irreducible. To compute I_2, we reason as follows. There are a total of 4 monic polynomials of degree 2; and a polynomial of degree two is either irreducible or a product of two (not necessarily distinct) irreducibles of degree 1. Since there are 2 irreducibles of degree 1, there are $\frac{2\cdot 3}{2\cdot 1} = 3$ ways to form a product of two of them, and so 3 *reducible* monic polynomials of degree 2. Thus $I_2 = 1$. To compute I_3, we note that there are 8 monic polynomials of degree 3, and a third degree polynomial must have a factorization into irreducibles of degrees (3), (2,1), or (1,1,1). The number of such products of irreducibles is I_3, $I_2 I_1$, and $\binom{I_1+2}{3}$, respectively. Thus

$$8 = I_3 + I_2 I_1 + \binom{I_1 + 2}{3},$$

but since we already know that $I_1 = 2$ and $I_2 = 1$, it follows that $I_3 = 2$. We could continue in this way to compute $I_4, I_5, \ldots$, but we can "automate" the process by introducing *generating functions.* (The following development assumes the reader to be somewhat familiar with the calculus of generating functions, and will omit some details.)

The idea is to define the generating function $M(z)$ as

$$M(z) = \sum_{n \geq 0} T_n z^n, \tag{6.3}$$

where T_n denotes the total number of monic polynomials of degree n over k, and to obtain two different analytic expression for $M(z)$, reflecting two different ways of counting T_n. The first way is simply to observe that since the coefficients of $1, x, \ldots, x^{n-1}$ can each assume q different values, there are exactly q^n monic polynomials of degree n. Thus

$$M(z) = \sum_{n \geq 0} q^n z^n = \frac{1}{(1-qz)}. \tag{6.4}$$

The second way is to use the fact that every monic polynomial of degree n factors uniquely into a product of monic irreducible polynomials. If we are given a *particular* monic irreducible polynomial $p(x)$ of degree d, then the number of degree n monic polynomials which are powers of $p(x)$ is 0 or 1 depending on whether n is a multiple of d or not. Therefore the generating function

$$1 + z^d + z^{2d} + z^{3d} + \cdots = \frac{1}{(1-z^d)}.$$

counts the monic polynomials which are powers of $p(x)$, in the sense that the coefficient of z^n is equal to the number of monic polynomials of degree n which are powers $p(x)$. Hence the infinite product

$$\prod_{d \geq 1} \left(\frac{1}{1-z^d} \right)^{I_d}$$

counts the monic polynomials which are products of powers of irreducible polynomials, i.e., all monic polynomials. Thus we have the following remarkable equation, that expresses the unique factorization theorem analytically:

$$\prod_{d \geq 1} \left(\frac{1}{1-z^d} \right)^{I_d} = \frac{1}{1-qz}. \tag{6.5}$$

Now we shall exploit (6.5), using calculus. If we take the derivative of the logarithm of both sides of (6.5), and multiply both sides by z, we obtain, after some algebra,

$$\sum_{d \geq 1} dI_d \left(\frac{z^d}{1-z^d} \right) = \frac{qz}{1-qz}. \tag{6.6}$$

The left side of (6.6) can be manipulated as follows:

$$\sum_{d\geq 1} dI_d\left(\frac{z^d}{1-z^d}\right) = \sum_{d\geq 1} dI_d \sum_{m\geq 1} z^{dm}$$
$$= \sum_{n\geq 1} z^n \sum_{d|n} dI_d.$$

On the other hand, the right side of (6.6) is plainly $\sum_{n\geq 1} z^n q^n$. Equating coefficients of z^n on both sides of (6.6), then, we get the result in Corollary 6.2 again.

Our next goal is to "solve" the equations of Theorem 5.6, Theorem 6.1 and Corollary 6.2 for $\phi(n)$, $V_d(x)$, and I_d, respectively. We shall see that all three of these equations can be solved by a general technique called *Möbius inversion.* We now pause to present a discussion of this important technique.

Let G be an Abelian group, and let us arbitrarily denote the group operation by "+" (rather than e.g. "·"). Suppose $a(1), a(2), \ldots$ and $b(1), b(2), \ldots$ are two sequences of elements in this group, and that they are related by

$$a(n) = \sum_{d|n} b(d), \qquad \text{for all } n \geq 1. \tag{6.7}$$

We shall consider the problem of *inverting* (6.7), i.e., of finding a formula that expresses the b's in terms of the a's.

Example 6.2. Let $G =$ the ordinary integers under the operation of ordinary addition. Let $a(n) = n$ and $b(d) = \phi(d)$. Then according to Theorem 5.6, (6.7) is satisfied. ∎

Example 6.3. Let $G =$ the set of rational functions over the finite field k, i.e., all expressions of the form

$$g = \frac{p(x)}{q(x)}, \qquad q(x) \neq 0,$$

where $p(x)$ and $q(x)$ are polynomials with coefficients in k. Let the group operation be multiplication, i.e.,

$$g_1 \cdot g_2 = \frac{p_1(x)p_2(x)}{q_1(x)q_2(x)},$$

if $g_i = p_i/q_i$. Then G is a commutative group. Now let

$$a(n) = x^{q^n} - x, \quad b(n) = V_n(x).$$

Then Theorem 6.1 says that (6.7) is satisfied. ∎

Example 6.4. Similarly, if $G =$ the integers Z under the operation of addition, (6.7) is satisfied by

$$a(n) = q^n, \quad b(n) = n \cdot I_n,$$

by Corollary 6.2. ∎

Example 6.5. Finally, take $G = Z$ again; $a(n) = n$, $b(n) = \phi(n)$. Then (6.7) is again satisfied—see Theorem 5.6. ∎

As a start, notice that (6.7) determines $b(n)$ uniquely in terms of the $a(n)$, since $b(1) = a(1)$ and for $n \geq 1$,

$$b(n) = a(n) - \sum_{\substack{d|n \\ d\neq n}} b(d) \tag{6.8}$$

By using (6.8) and a little patience, one can discover, e.g.,

$$\begin{aligned} b(1) &= a(1) \\ b(2) &= a(2) - a(1) \\ b(3) &= a(3) - a(1) \\ b(4) &= a(4) - a(2) \\ b(5) &= a(5) - a(1) \end{aligned}$$

$$b(6) = a(6) - a(3) - a(2) + a(1)$$

$$\vdots$$

What is needed, however, is a general formula.

We will now show that if we can invert (6.7) in the special case where the $a(n)$ sequence is $(1, 0, 0, 0, \ldots)$, we can invert any sequence. Thus we define the sequence $\mu(n)$ to be the "inverse" of the sequence $(1, 0, 0, 0, \ldots)$, i.e.,

$$\begin{aligned} \mu(1) &= 1, \\ \sum_{d|n} \mu(d) &= 0, \quad \text{for } n > 1. \end{aligned} \tag{6.9}$$

We don't know much about this sequence yet, but as we observed above, it is uniquely determined, and we have $\mu(1) = 1$, $\mu(2) = -1$, $\mu(3) = -1$, $\mu(4) = 0$, $\mu(5) = -1$, $\mu(6) = 1$, etc. Anyway, the inversion formula we're looking for is this:

$$b(n) = \sum_{d|n} a(d)\mu(n/d). \tag{6.10}$$

To prove (6.10), we define the sequence $b'(n)$ by

$$b'(n) = \sum_{d|n} a(d)\mu(n/d),$$

and show that $(a(n))$ and $(b'(n))$ are related by (6.7). Here we go:

$$\begin{aligned} \sum_{d|n} b'(d) &= \sum_{d|n} \sum_{e|d} a(e)\mu(d/e) \\ &= \sum_{e|n} a(e) \sum_{d:e|d|n} \mu(d/e) \\ &= \sum_{e|n} a(e) \sum_{f|(n/e)} \mu(f) \end{aligned}$$

$$= a(n) \text{ (by (6.9))}.$$

Thus both $b(n)$ (by assumption) and $b'(n)$ (by the above calculation) satisfy (6.7). But since the b's are uniquely determined by the a's (see (6.8)), it follows that $b'(n) = b(n)$, i.e., (6.10) holds for *any* set of $(a(n))$, $(b(n))$ related by (6.7). ∎

All that remains is to give a more practical method for computing the basic numbers $\mu(n)$. In fact there is an explicit formula for $\mu(n)$ in terms of the factorization of n. Let $n = p_1^{e1} p_2^{e2} \cdots p_m^{em}$ be the factorization of n into distinct prime powers. Then we assert that

$$\mu(n) = \begin{cases} 1, & \text{if } n = 1; \\ 0, & \text{if any } e_i \text{ is} \geq 2 \\ (-1)^m, & \text{otherwise.} \end{cases} \tag{6.11}$$

In other words, $\mu(n) = 0$ if n is divisible by the square of any prime, $\mu(n) = 1$ if n is the product of an even number of distinct primes, and $\mu(n) = -1$ if n is the product of an odd number of distinct primes. To prove that the μ defined in (6.9) is the same as the μ defined in (6.11), we merely verify that (6.9) holds, if μ is defined as in (6.11). To do this, let $x_1, x_2, \ldots, x_m$ be indeterminates, and for each divisor d of n define $x(d)$ as follows:

$$x(d) = \prod_{p_i | d} x_i.$$

Thus $x(1) = 1$, $x(p_1) = x_1$, $x(p_1 p_2^2) = x_1 x_2$, and so on. We now expand the product $\prod_{i=1}^{m} (1 - x_i)$.

$$\prod_{i=1}^{m} (1 - x_i) = \sum_{d|n} \mu(d) x(d), \tag{6.12}$$

where in (6.12) we are using the definition (6.11) of $\mu(n)$. If we now substitute 1 for each of the x_i's in (6.12), we obtain

$$\sum_{d|n} \mu(d) = \begin{cases} 1, & \text{if } n = 1; \\ 0, & \text{if } n > 1. \end{cases}$$

Since this agrees with the original definition (6.9) of $\mu(n)$, the formula (6.11) must be correct. Summarizing, we have proved

Theorem 6.3. *If $(a(n)), (b(n))$ are two sequences of elements in a commutative group G which satisfy (6.7), then (6.7) may be inverted to give*

$$b(n) = \sum_{d|n} a(d)\mu(n/d) = \sum_{d|n} \mu(d)a(n/d),$$

where the function μ is given by (6.11). ∎

The formula which expresses $b(n)$ in terms of $a(n)$ is called the *Möbius Inversion Formula*, and the function $\mu(n)$ is called the *Möbius* function. We will see many applications of the Möbius inversion in the sequel.

Our first application of Möbius inversion will be to combine Theorems 5.6 and 6.3 to get a computationally convenient formula for the Euler function $\phi(n)$. Theorem 5.6 says that

$$n = \sum_{d|n} \phi(d).$$

Applying Theorem 6.3 with $a(n) = n$ and $b(d) = \phi(d)$, we get

$$\phi(n) = \sum_{d|n} \mu(d) \cdot \tfrac{n}{d}$$

i.e.,

$$\phi(n) = n \sum_{d|n} \frac{\mu(d)}{d}. \tag{6.13}$$

But we can say even more: if

$$n = p_1^{e_1} p_2^{e_2} \cdots p_m^{e_m},$$

is the factorization of n into distinct prime powers, then applying (6.12) with $x_i = 1/p_i$, we find that

$$\sum_{d|n} \frac{\mu(d)}{d} = \prod_{i=1}^{m}(1 - \tfrac{1}{p_i}). \tag{6.14}$$

Combining (6.13) and (6.14), we get the following nice formulas for $\phi(n)$:

$$\phi(n) = n \cdot \prod_{p|n}(1 - \tfrac{1}{p}) \tag{6.15}$$

$$= \prod_{i=1}^{m} p_i^{e_i - 1}(p_i - 1), \tag{6.16}$$

where in the product in (6.15) the symbol $p \mid n$ denotes "all distinct prime divisors of n."

Example 6.6. Let $n = 180 = 2^2 \cdot 3^2 \cdot 5$. Then by (6.15)

$$\begin{aligned}\phi(180) &= 180(1 - \tfrac{1}{2})(1 - \tfrac{1}{3})(1 - \tfrac{1}{5}) \\ &= 48.\end{aligned}$$

Similarly by (6.16) if $n = 2646 = 2 \cdot 3^3 \cdot 7^2$

$$\phi(2646) = 1 \cdot 3^2(2) \cdot 7(6) = 756.$$ ■

As our next application of Möbius inversion, we'll use Corollary 6.2 to find a formula for the number of irreducible polynomials of a given degree over a given finite field. Corollary 6.2 says that

$$q^n = \sum_{d|n} dI_d.$$

This is a special case of (6.7) with $a(n) = q^n$, $b(n) = nI_n$ (the underlying group being the integers, under ordinary addition). Thus by Theorem 6.3, we have

$$I_n = \tfrac{1}{n}\sum_{d|n}\mu(d)q^{n/d}, \tag{6.17}$$

which is an explicit formula for the number of irreducible polynomials of degree n over a finite field with q elements. Since the dominant term in (6.17) occurs for $d = 1$, we get the estimate

$$I_n \approx \tfrac{q^n}{n}, \tag{6.18}$$

which can be expected to hold for large n if q is held fixed. Since the total number of (monic) polynomials of degree n is just q^n, if we choose a polynomial of degree n at random, the probability of its being irreducible is about $1/n$. This is interesting but it doesn't prove that there exists an irreducible polynomial of any specific degree.

Let's look at (6.17) more closely for the first few values of n:

$$\begin{aligned} I_1 &= q \\ I_2 &= \tfrac{1}{2}(q^2 - q) \\ I_3 &= \tfrac{1}{3}(q^3 - q) \\ I_4 &= \tfrac{1}{4}(q^4 - q^2) \\ I_5 &= \tfrac{1}{5}(q^5 - q) \\ I_6 &= \tfrac{1}{6}(q^6 - q^3 - q^2 + q), \quad \text{etc.} \end{aligned}$$

From this, we can see that I_n is never zero. Why? Consider I_6. From the above calculations,

$$\frac{6I_6}{q} = q^5 - q^2 - q + 1,$$

which, being an integer congruent to 1 $(\bmod q)$, cannot be zero. Thus $I_6 \neq 0$, and so there are always irreducible sixth degree polynomials. In general, (6.17) shows that nI_n is a sum of signed but distinct powers of q, and such a sum can never be zero. Hence $I_n > 0$ for all n and q. We thus have proved the following theorem.

Theorem 6.4. *If there exists a field F with q elements, then for all $n \geq 1$, there exists at least one irreducible polynomial of degree n over F.* ∎

One immediate consequence of Theorem 6.4 is that *for any positive integer q of the form $q = p^m$, where p is a prime, there exists a finite field of order q.* The proof should by now be obvious: let $f(x)$ be a polynomial in $F_p[x]$ irreducible of degree m. Then

$$F = F_p[x] \pmod{f(x)} \tag{6.19}$$

is a finite field with q elements. This completely settles the question "for which integers q does a finite field with q elements exist?"; such a field exists if and only if q is a power of a prime. (See Theorem 5.1.)

However, we have not yet settled the question as to whether or not there might be *more than one* field of a given order. A little care must be exercised here; for as we saw in Example 4.5 one field can appear in many different disguises if we change the basis around. We shall now show that any two finite field of the same order must be disguised versions of each other, i.e. *isomorphic*.

Thus let F be a particular field of order p^m, say the one constructed in Eq. (6.19); and let E be some other field of order p^m. We want to show that E and F are isomorphic. E can be viewed as an m–dimensional vector space over F_p (see proof of Theorem 5.1); conversely, F_p can be viewed as a subfield of E.

Now consider the factorization of $x^{p^m} - x$ over the two fields E and F_p. On one hand, according to Theorem 6.1, $x^{p^m} - x$ factors over F_p into the product of *all* irreducible monic polynomials whose degree divides m. In particular, the polynomial $f(x)$ of (6.19) divides $x^{p^m} - x$:

$$x^{p^m} - x = f(x)J(x), \tag{6.20}$$

for some $J(x) \in F_p[x]$. On the other hand, again according to Theorem 6.1,

$x^{p^m} - x$ factors over E into a product of linear factors:

$$x^{p^m} - x = \prod_{\beta \in E} (x - \beta). \tag{6.21}$$

Comparing (6.20) and (6.21), we see that for every $\beta \in E$, $f(\beta) \cdot J(\beta) = 0$. But J, having degree $p^m - m$ can have at most $p^m - m$ zeros. Thus *there are m elements in E such that* $f(\beta) = 0$, i.e., f factors completely into linear factors in E.

Now let α be any root of $f(x) = 0$ in E. The m elements

$$\{1, \alpha, \alpha^2, \ldots, \alpha^{m-1}\}$$

are linearly independent. For if say $\sum \lambda_k \alpha^k = 0$, then $\lambda(x) = \lambda_0 + \lambda_1 x + \cdots + \lambda_{m-1} x^{m-1}$ would be a polynomial in $F_p[x]$ having α as a root. Then if $g(x) = \gcd(\lambda(x), f(x))$, we would have $g(\alpha) = 0$. But $f(x)$ is *irreducible* over F_p, which implies that $g(x) = 1$, a contradiction.

Since $\{1, \alpha, \ldots, \alpha^{m-1}\}$ are linearly independent, it follows that every element $\beta \in E$ has a unique expression of the form

$$\beta = \lambda_0 + \lambda_1 \alpha + \cdots + \lambda_{m-1} \alpha^{m-1}, \qquad \lambda_i \in F_p,$$

i.e., $\{1, \alpha, \ldots, \alpha^{m-1}\}$ is a *basis* for the vector space E over the field F_p. We claim that when the elements of E are expressed with respect to this basis, the arithmetic in E is identical to the arithmetic in the field F constructed as in (6.19).

Given two elements $\beta, \beta' \in E$, expressed with respect to the specific basis $\{1, \alpha, \ldots, \alpha^{m-1}\}$ how do we multiply them? Let

$$\beta = \sum_{k=0}^{m-1} \lambda_k \alpha^k$$

$$\beta' = \sum_{k=0}^{m-1} \lambda'_k \alpha^k$$

be the expansions of β and β', and let

$$\beta(x) = \sum_{k=0}^{m-1} \lambda_k x^k$$

$$\beta'(x) = \sum_{k=0}^{m-1} \lambda'_k x^k$$

be the corresponding polynomials. We now multiply the polynomials $\beta(x)$ and $\beta'(x)$ (mod $f(x)$):

$$\beta(x)\beta' \equiv r(x) \pmod{f(x)}$$

where

$$r(x) = r_0 + r_1 x + \cdots + r_{m-1} x^{m-1}, \qquad r_i \in F_p.$$

Then, because $f(\alpha) = 0$, it follows that

$$\beta \cdot \beta' = r$$

where $r = r_0 + r_1\alpha + \cdots + r_{m-1}\alpha^{m-1}$. But this is exactly the rule for multiplication in $F = F_p[x] \pmod{f(x)}$! Thus when the elements of E are expressed with respect to the basis $\{1, \alpha, \ldots, \alpha^{m-1}\}$, the field is indistinguishable from the field constructed in (6.19). These results we summarize in a theorem:

Theorem 6.5. *For any prime power p^m, there is (up to isomorphism) one and only one finite field: and it may be constructed as in (6.19).*

Notation. *The finite field of order p^m is denoted by the symbol $GF(p^m)$ (GF = Galois Field).*

Example 6.7. Let $F = F_2[x] \pmod{x^4+x^3+x^2+x+1}$, $E = F_2[x] \pmod{x^4+x+1}$. In F, let $\beta = \overline{x}$; and in E let $\alpha = \overline{x}$. A simple calculation (Example 5.6) shows that in E, α^3 has minimal polynomial $x^4+x^3+x^2+x+1$, and so α^3 is a root of the defining polynomial of F. Thus, if we choose the basis $1, \alpha^3, \alpha^6, \alpha^9$

for E, the rule for multiplying vectors will be the same as in F with respect to the basis $1, \beta, \beta^2, \beta^3$. ∎

The final topic we'll consider in this chapter is the topic of *subfields*. A subfield k of a field F is a subset of F which is itself a field. As a familiar example, we note that the field of real numbers is a subfield of the field of complex numbers. But of course we're only interested in finite fields. Since Theorem 6.5 tells us that the only field of order p^n is $GF(p^n)$, we ask the following question: What subfields does $GF(p^n)$ contain?

Theorem 6.6. *For every divisor d of n, $GF(p^n)$ contains exactly one subfield isomorphic to $GF(p^d)$; and $GF(p^n)$ has no other subfields.*

Proof: Let $F = GF(p^n)$, and let k be a subfield of F. By Theorem 6.5, $k \sim GF(p^d)$ for some $d < n$. Now by Lemma 5.10, every element $\alpha \in k$ must satisfy the equation $x^{p^d} - x = 0$. But also every element of k when viewed as an element of F must satisfy $x^{p^n} - x = 0$. Thus every root of $x^{p^d} - x = 0$ is also a root of $x^{p^n} - x = 0$, and so $(x^{p^d} - x) \mid (x^{p^n} - x)$. By the Corollary to Theorem 2.3, this implies $d \mid n$. We conclude that if $GF(p^n)$ has a subfield isomorphic to $GF(p^d)$, then $d \mid n$.

Now suppose $d \mid n$. If F has a subfield isomorphic to $GF(p^d)$, it must be the set $k = \{\alpha \in F : \alpha^{p^d} = \alpha\}$. We shall show that k contains exactly p^d elements, and is a field. Since $x^{p^d} - x \mid x^{p^n} - x$, $x^{p^n} - x = \prod(x - \alpha : \alpha \in F)$, it follows that $x^{p^d} - x = 0$ has p^d solutions in F, i.e., k contains p^d elements. We claim k is a field. To verify this we must show that k is closed under addition, multiplication, and inversion. Thus let $\alpha, \beta \in k$, and note the following.

$$\begin{aligned}
(\alpha + \beta)^{p^d} &= \alpha^{p^d} + \beta^{p^d} = \alpha + \beta \\
(\alpha\beta)^{p^d} &= \alpha^{p^d} \cdot \beta^{p^d} = \alpha \cdot \beta \\
(\alpha^{-1})^{p^d} &= (\alpha^{p^d})^{-1} = \alpha^{-1}.
\end{aligned}$$

Thus k, being closed under these operations, must be a field, and having p^d elements, by Theorem 6.5 it is (isomorphic to) $GF(p^d)$. This completes the proof of Theorem 6.6. ∎

Example 6.8. Let $n = 12$. Then the lattice of divisors of the integer 12 looks like this:

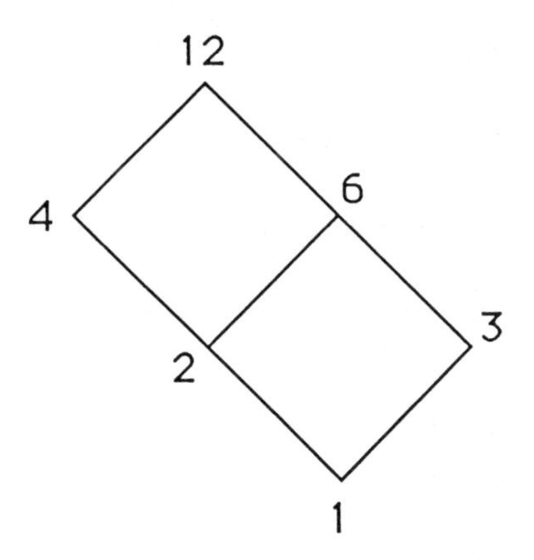

It follows then from Theorem 6.6 that the subfield structure of $GF(p^{12})$ looks like this:

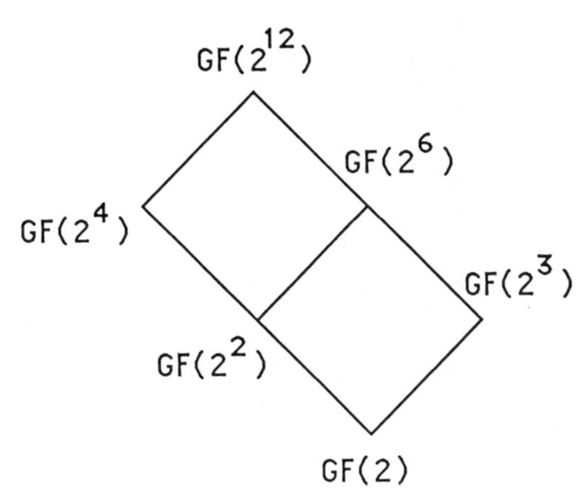

(Implicit in these diagrams is the (slight) generalization of Theorem 6.6 that says

$$GF(p^m) \cap GF(p^n) = GF(p^{\gcd(m,n)}), \tag{6.22}$$

whose proof we leave to the reader.) ∎

Problems for Chapter 6.

1. (Formal Derivatives). If $P(x) = P_0 + P_1x + \cdots + P_nx^n$ is a polynomial over a field F its *formal derivative* $P'(x)$ is defined by

$$P'(x) = P_1 + 2P_2x + \cdots + nP_nx^{n-1}.$$

From this definition, and without the use of limits, deduce the following facts.

a. $(P+Q)' = P' + Q'$.
b. $(PQ)' = PQ' + P'Q$.
c. $(P^m)' = mP^{m-1}P'$.
d. Finally show that if

$$P(x) = P_1(x)^{e_1}P_2(x)^{e_2}\cdots P_m(x)^{e_m}$$

is the factorization of $P(x)$ into powers of distinct irreducible factors, then

$$\frac{P(x)}{\gcd(P(x), P'(x))} = P_1(x)P_2(x)\cdots P_m(x).$$

As a corollary, show that $P(x)$ is squarefree if and only if it is relatively prime to its formal derivative.

2. Calculate $\mu(11)$, $\mu(101)$, and $\mu(1001)$,

3. Calculate $\phi(11)$, $\phi(101)$, and $\phi(1001)$.

4. Show that

$$\sum_{d^2|n} \mu(d) = |\mu(n)|.$$

5. Evaluate the sum $\sum_{d|n} |\mu(d)|$.

6. Use formula (6.16) to do the following.

a. Prove that

$$\limsup_{n\to\infty} \frac{\phi(n)}{n} = 1$$

$$\liminf_{n\to\infty} \frac{\phi(n)}{n} = 0.$$

b. Find the smallest n such that $\phi(n)/n > 0.99$.

c. Find the smallest n such that $\phi(n)/n < 0.1$.

7. In the Proof of Theorem 6.5, we didn't check that the *addition* was the same in both fields. Is this a problem?

8. Prove (6.22), using Theorem 6.6.

9. Draw the lattice of subfields for $GF(p^{16})$, $GF(p^{20})$, and $GF(p^{24})$.

Chapter 7

Factoring Polynomials over Finite Fields

We begin this chapter by considering another application of Möbius inversion, this time to the result of Theorem 6.1, viz.

$$x^{q^n} - x = \prod_{d|n} V_d(x),$$

where $V_d(x)$ denotes the product of all monic irreducible polynomials of degree d over $GF(q)$. We recall (see Example 6.3) that here the underlying group is the set of rational functions $\{p(x)/q(x) : p(x), q(x) \in k[x], p(x) \neq 0, q(x) \neq 0\}$, and the group operation is multiplication. When the group is written multiplicatively, Theorem 6.3 becomes

$$b_n = \prod_{d|n} a_d^{\mu(n/d)}, \tag{7.1}$$

where $a^0 = 1$ (the group's identity), $a^{+1} = a$, $a^{-1} =$ the inverse of a. Thus combining (7.1) with Theorem 6.1, we obtain

$$V_n(x) = \prod_{d|n} (x^{q^d} - x)^{\mu(n/d)}. \tag{7.2}$$

Example 7.1. Let $q = 2, n = 6$. Then according to (7.2),

$$\begin{aligned} V_6(x) &= \frac{(x^2 - x)(x^{64} - x)}{(x^8 - x)(x^4 - x)} \\ &= \frac{(x-1)(x^{63}-1)}{(x^7-1)(x^3-1)} \\ &= \text{a complicated polynomial of degree 54.} \end{aligned}$$

This complicated polynomial is by Theorem 6.1 the product of the 9 irreducible monic polynomials of degree 6. But how to find those polynomials individually? This problem is a little too hard for us right now, but we can make preliminary progress by studying the *cyclotomic polynomials.* ∎

We begin this new topic, ironically, by considering the infinite field C of complex numbers. Let n be a positive integer, and let $\varsigma = \exp(2\pi i/n)$ a primitive complex nth root of unity. Then over C we have the factorization

$$x^n - 1 = \prod_{j=0}^{n-1} (x - \varsigma^j). \tag{7.3}$$

As we observed previously (see Lemma 5.4), the order of ς^j depends on $\gcd(j, n)$, and for every divisor d of n there are exactly $\phi(d)$ powers of ς which have order d. We define the dth cyclotomic polynomial $\Phi_d(x)$ to be the monic polynomial of degree $\phi(d)$ which has as its roots those powers of ς which have order d:

$$\Phi_d(x) = \prod \{(x - \varsigma^j) : 0 \le j \le n-1, \gcd(j, n) = n/d\}. \tag{7.4}$$

It follows from (7.3) and (7.4) that

$$x^n - 1 = \prod_{d|n} \Phi_d(x), \tag{7.5}$$

and hence by the Möbius inversion formula (7.2) that for all $n \geq 1$,

$$\Phi_n(x) = \prod_{d|n} (x^d - 1)^{\mu(n/d)}. \tag{7.6}$$

Now (7.6) leads us to a surprising observation. It shows that the coefficients of $\Phi_n(x)$ are *integers*, despite the appearance of complex numbers in the definition (7.4). This is because in (7.6) we have an expression for $\Phi_n(x)$ as a quotient of two monic polynomials with integer coefficients. (The numerator in this quotient is the product of all the terms $x^d - 1$ for which $\mu(n/d) = +1$, and the denominator is the product of all the terms for which $\mu(n/d) = -1$.) But the process of dividing one polynomial with integer coefficients by a *monic* polynomial with integer coefficients can only result in integer coefficients in the quotient!

For example, consider $\Phi_{18}(x)$. By (7.6) we have

$$\Phi_{18}(x) = \frac{(x^{18} - 1)(x^3 - 1)}{(x^9 - 1)(x^6 - 1)}.$$

We could therefore express this as the quotient of a degree 21 polynomial and a degree 15 polynomial, and then do the long division, but we prefer the following method, due to Elwyn Berlekamp. Now $\Phi_{18}(x)$ will have degree $\phi(18) = 6$, and so by considering these polynomials $\pmod{x^7}$ we won't disturb anything:

$$\begin{aligned}\Phi_{18}(x) &\equiv \frac{(1 - x^3)}{(1 - x^6)} \pmod{x^7} \\ &\equiv (1 - x^3)(1 + x^6) \pmod{x^7} \\ &\equiv (1 - x^3 + x^6) \pmod{x^7}.\end{aligned}$$

Thus $\Phi_{18}(x) = x^6 - x^3 + 1$.

Here is a short table of the first few cyclotomic polynomials:

n	$\Phi_n(x)$
1	$x-1$
2	$x+1$
3	x^2+x+1
4	x^2+1
5	$x^4+x^3+x^2+x+1$
6	x^2-x+1
7	$x^6+x^5+x^4+x^3+x^2+x+1$
8	x^4+1
9	x^6+x^3+1

Our concern is with finite fields, not the complex numbers. But the factorization (7.5) remains true over any field, since the coefficients of Φ_n are integers. (Of course in a field of characteristic p, the integer coefficients of $\Phi_n(x)$ must be interpreted as elements in the prime field F_p.)

Let us now return to Example 7.1, where we computed

$$V_6(x) = \frac{(x-1)(x^{63}-1)}{(x^7-1)(x^3-1)}.$$

In view of (7.5), however, we have

$$V_6(x) = \frac{\Phi_1 \cdot \Phi_1 \cdot \Phi_3 \cdot \Phi_7 \cdot \Phi_9 \cdot \Phi_{21} \cdot \Phi_{63}}{\Phi_1 \cdot \Phi_7 \cdot \Phi_1 \cdot \Phi_3}$$

$$= \Phi_9 \cdot \Phi_{21} \cdot \Phi_{63},$$

where Φ_9 has degree $\phi(9) = 6$, Φ_{21} has degree $\phi(21) = 12$, and Φ_{63} has degree $\phi(63) = 36$. We conclude that over the field $GF(2)$ Φ_9 (degree 6) is irreducible; Φ_{21} (degree 12) is the product of two distinct irreducibles of degree 6; and Φ_{63} (degree 36) is the product of 6 irreducibles of degree 6. In particular,

$$\Phi_9 = \frac{(x^9-1)}{(x^3-1)} = x^6+x^3+1$$

must be irreducible over $GF(2)$. To complete the factorization of V_6 into irreducible polynomials, however, we must learn how to factor the cyclotomic polynomials.

It is known that over the ordinary field of rational numbers, $\Phi(x)$ is irreducible for all $n \geq 1$. However, over finite fields, the situation is much more interesting. For example, consider $\Phi_4(x) = x^2 + 1$. This polynomial does not factor over the rationals, and yet we have

$$\begin{aligned} x^2 + 1 &= (x+1)^2 && \text{over } GF(2) \\ &= (\text{irreducible}) && \text{over } GF(3) \\ &= (x+1)^2 && \text{over } GF(4) \\ &= (x-2)(x-3) && \text{over } GF(5) \\ &= (\text{irreducible}) && \text{over } GF(7) \\ &\ \ \vdots \end{aligned}$$

So now let's consider the general problem of how $\Phi_n(x)$ factors over a given finite field. The next theorem will prove to be very useful, though it doesn't tell the whole story.

Theorem 7.1. *Suppose that p is a prime and that $\gcd(p,n) = 1$. Then for $k \geq 1$*

(*a*) $\Phi_{np^k}(x) = \Phi_{np}(x^{p^{k-1}})$

(*b*) $\Phi_{np^k}(x) = \dfrac{\Phi_n(x^{p^k})}{\Phi_n(x^{p^{k-1}})}$

(*c*) $\Phi_{np^k}(x) = \Phi_n(x)^{p^k - p^{k-1}}$ *in a field of characteristic p.*

Proof: By (7.6) (after interchanging d and n/d),

$$\Phi_{np^k}(x) = \prod_{d|np^k} (x^{np^k/d} - 1)^{\mu(d)}.$$

But since $\mu(d) = 0$ if $p^2 \mid d$, in fact we have

$$\Phi_{np^k}(x) = \prod_{d|np} (x^{np^k/d} - 1)^{\mu(d)} \tag{7.7}$$
$$= \Phi_{np}(x^{p^{k-1}}).$$

This proves (a). Let us now partition the divisors d of np which appear in (7.7) into two sets: the divisors which *are* and *are not* divisible by p. The divisors *not* divisible by p contribute

$$\prod_{d|n}(x^{np^k/d} - 1)^{\mu(d)} = \Phi_n(x^{p^k}).$$

The divisors which *are* divisible by p contribute

$$\prod_{d|n}(x^{np^{k-1}/d} - 1)^{\mu(pd)} = \prod_{d|n}(x^{np^{k-1}/d} - 1)^{-\mu(d)}$$
$$= \Phi_n(x^{p^{k-1}})^{-1}.$$

This proves (b). Finally, by Lemma 5.12, $f(x^p) = f(x)^p$ in a field of characteristic p, and so it follows from (b) that in such a field

$$\Phi_{np^k}(x) \equiv \frac{\Phi_n(x)^{p^k}}{\Phi_n(x)^{p^{k-1}}} = \Phi_n(x)^{(p^k - p^{k-1})}.$$

This proves (c). ∎

Example 7.2. Consider $\Phi_{72}(x) = \Phi_{8\cdot 9}(x)$. By two applications of Theorem 7.1a we have

$$\Phi_{72}(x) = \Phi_6(x^{12}) = x^{24} - x^{12} + 1.$$

By Theorem 7.1b we have

$$\begin{aligned}\Phi_{72}(x) &= \frac{\Phi_8(x^9)}{\Phi_8(x^3)}\\ &= \frac{(x^{36}+1)}{(x^{12}+1)}\\ &= x^{24} - x^{12} + 1.\end{aligned}$$

Finally, by Theorem 7.1c, we have in any field of characteristic 3 that

$$\begin{aligned}\Phi_{72}(x) &= \Phi_8(x)^6\\ &= (x^4+1)^{3\cdot 2}\\ &= (x^{12}+1)^2\\ &= x^{24} + 2x^{12} + 1\end{aligned}$$

Similarly, in any field of characteristic 2, we have

$$\begin{aligned}\Phi_4(x) &= \Phi_2(x)^2 = (x+1)^2\\ \Phi_6(x) &= \Phi_3(x) = x^2 + x + 1\\ \Phi_8(x) &= \Phi_2^4 = (x+1)^4\\ &\vdots\end{aligned}$$

∎

In view of Theorem 7.1c, when factoring $\Phi_n(x)$ over a field of characteristic p, we may safely assume that $\gcd(n,p) = 1$. Having made this assumption, we are assured that there is a least integer m such that

$$q^m = 1 \pmod{n}, \tag{7.8}$$

$q = p^e$ being the order of k. By Theorem 6.5, there is a field $F \simeq GF(q^m)$ of order q^m. This field will contain by (7.8) and Theorem 5.7 an element α of order n. Then by reasoning exactly like that which led to (7.6), we can conclude that for every divisor d of n,

$$\begin{aligned}\Phi_d(x) &= \prod\{x - \alpha^j : 0 \le j \le n-1 \quad \text{and} \quad \gcd(j,n) = n/d\} \\ &= \prod\{x - \beta : \operatorname{ord}(\beta) = d\}.\end{aligned}$$

What this means is that in the big field F, $\Phi_d(x)$ factors completely into linear factors. To see how it factors in the small field k, we will use what we already know about minimal polynomials.

The minimal polynomial of α has as roots $\alpha, \alpha^q, \alpha^{q^2}$, etc., and indeed by Theorem 5.13 there are exactly d conjugates, where d is the smallest integer such that $q^d \equiv 1 \pmod{n}$. But we have defined this number to be m in Eq. (7.8). We conclude that $\Phi_n(x)$ has one irreducible factor of degree m, viz., the minimal polynomial of α.

What about the minimal polynomial of the other roots of $\Phi_n(x)$? They all have, by definition, order precisely n, and so by the same argument given above, they all have exactly d conjugates. Hence we have:

Theorem 7.2. *Let $p \not| \, n, q = p^e$. Then over the field $GF(q)$, the cyclotomic polynomial $\Phi_n(x)$ factors into the product of $\phi(n)/m$ irreducible factors of degree m, where m is determined as the least integer such that (7.8) holds.*

Example 7.3. Let's consider the factorization of $\Phi_7(x)$ over $GF(2)$. We have $2^3 \equiv 1 \pmod 7$, and $\phi(7) = 6$, and so by Theorem 7.2 we know that Φ_7 factors into two irreducible cubics. In fact, if α denotes an element of order 7 in $GF(2^3)$, these two irreducibles are

$$\begin{aligned}f_1(x) &= (x-\alpha)(x-\alpha^2)(x-\alpha^4) \\ f_3(x) &= (x-\alpha^3)(x-\alpha^6)(x-\alpha^5).\end{aligned}$$

If α satisfies the cubic equation $\alpha^3 = \alpha + 1$, then it isn't hard to calculate that

$$\begin{aligned}f_1(x) &= x^3 + x + 1 \\ f_3(x) &= x^3 + x^2 + 1.\end{aligned}$$

■

Example 7.4. Next let us determine the shape of the factorization of $\Phi_{180}(x)$ over $GF(3)$. We have $180 = 2^2 \cdot 3^2 \cdot 5$, so according to Theorem 7.1c,

$$\Phi_{180}(x) = (\Phi_{20}(x))^6.$$

But how does $\Phi_{20}(x)$ factor? We have $3^4 \equiv 1 \pmod{20}$, so by Theorem 7.2, Φ_{20} factors into $\phi(20)/4 = 2$ irreducible factors, each of degree 4. In fact, if α denotes an element of order 20 in $GF(3^4)$, these two irreducible factors are

$$f_1(x) = (x-\alpha)(x-\alpha^3)(x-\alpha^9)(x-\alpha^7)$$

$$f_{11}(x) = (x-\alpha^{11})(x-\alpha^{13})(x-\alpha^{19})(x-\alpha^{17}),$$

and if we had an explicit realization of $GF(81)$ we could calculate $f(x)$ and $f_{11}(x)$ explicitly. In any event, our conclusion is that $\Phi_{180}(x)$ factors into 12 irreducible factors of degree 4. ∎

Our next goal is to discover techniques that will allow us to find the factors of $\Phi_n(x)$ over $GF(q)$ *explicitly*; but before we do this, let's briefly discuss the simplest possible situation—when $\Phi_n(x)$ is already irreducible. By Theorem 7.2 this will happen only if

$$\text{(7.9)} \qquad q^{\phi(n)} \equiv 1 \pmod{n}, \quad q^k \not\equiv 1 \pmod{n} \qquad \text{if } k < \phi(n).$$

Now the set of residues (mod n) which are relatively prime to n forms a multiplicative group of order $\phi(n)$. What (7.9) says is that this group is *cyclic*, and that q is a generator for this group. In this case, q is called a *primitive root* (mod n). It is shown in texts on number theory that the only values of n for which the group of residues (mod n) is cyclic are

$$\text{(7.10)} \qquad n = 1, 2, 4, p^s, 2p^s,$$

where p is any odd prime, i.e., the n's in the sequence

$$1, 2, 3, 4, 5, 6, 7, 9, 10, 11, 13, 14, 17, 18, \ldots.$$

Thus if n is not of this form $\Phi_n(x)$ cannot be irreducible over any finite field. On the other hand if n is of the form (7.10), $\Phi_n(x)$ will be irreducible over $GF(q)$ if q is a generator for the group of residues which are relatively prime to n. For example, let $n = 7$. The primitive roots (mod 7) are 3 and 5. Hence $\Phi_7(x) = x^6 + x^5 + x^4 + x^3 + x^2 + x + 1$ is irreducible over $GF(q)$ if and only if

$$q \equiv 3 \pmod 7 \qquad \text{or} \qquad q \equiv 5 \pmod 7.$$

For example, $q = 3, 5, 17, 19, 31, 59, 61$, etc., all correspond to $GF(q)$'s where $\Phi_7(x)$ is irreducible.

On the other hand, $\phi(8) = 4$, but there is no element of order $4 \pmod 8$ and so $\Phi_8(x) = x^4 + 1$ is reducible (mod p) for every p. This is in spite of the fact that $\Phi_8(x)$ is irreducible over the rationals.

Finally, note that $\Phi_n(x)$ is irreducible over $GF(2)$ if and only if 2 is a primitive root (mod n), i.e., if 2 has order $\phi(n) \pmod n$. These n are the primes in the sequence

$$n = 3, 5, 9, 11, 13, 19, 25, 29, \dots .$$

This gives us explicit irreducible polynomials of degrees

$$d = 2, 4, 6, 10, 12, 18, 20, 28, \text{ etc.}$$

For other degrees, we aren't so lucky and $\Phi_n(x)$ is reducible. In these cases we naturally wish to find the irreducible factors. We now present a practical way to factor not only the cyclotomic polynomial $\Phi_n(x)$, but indeed *any* polynomial $f(x)$ with coefficients in the finite field $k = GF(q)$. This algorithm is due to Berlekamp, and is based on the following theorem.

Theorem 7.3. *Let $f(x)$ be a monic polynomial of degree n with coefficients in $GF(q) = k$. Then if $h(x) \in k[x]$ is such that*

$$h(x)^q \equiv h(x) \pmod{f(x)}, \tag{7.11}$$

then

$$f(x) = \prod_{s \in k} \gcd(f(x), h(x) - s). \tag{7.12}$$

Proof: First note that because of Theorem 6.1, which implies that $y^q - y = \prod\{y - s : s \in k\}$, we have, for any polynomial $h(x)$,

$$h(x)^q - h(x) = \prod_{s \in k} (h(x) - s).$$

Thus the hypothesis (7.11) can be written as follows:

$$f(x) \mid \prod_{s \in k} (h(x) - s).$$

This clearly implies

$$f(x) \mid \prod_{s \in k} \gcd(f(x), h(x) - s). \tag{7.13}$$

On the other hand, for each $s \in k$ we have

$$\gcd(f(x), h(x) - s) \mid f(x). \tag{7.14}$$

Furthermore, if $s_1 \neq s_2$, $h(x) - s_1$ and $h(x) - s_2$ are relatively prime. This implies that $\gcd(f(x), h(x) - s_1)$ and $\gcd(f(x), h(x) - s_2)$ are relatively prime, and hence from (7.14) we get

$$\prod_{s \in k} \gcd(f(x), h(x) - s) \mid f(x). \tag{7.15}$$

Together (7.13) and (7.15) imply Theorem 7.3 (recall the convention that gcd's are monic). ∎

Notice that if there exists an element $s \in k$ such that $h(x) \equiv s \pmod{f(x)}$, the factorization given by Theorem 7.3 is trivial, in the sense that one of the factors will be $f(x)$ and the others will all be 1. The next theorem will show that if $f(x)$ is divisible by two or more distinct irreducible polynomials, there exists a polynomial $h(x)$ satisfying (7.11) such that the factorization (7.12) is *non-trivial.*

The set of polynomials (mod $f(x)$), i.e., the ring $k[x]$ (mod $f(x)$) can be viewed as an n dimensional vector space $V(f)$ over k. We can take $\{1, x, x^2, \ldots, x^{n-1}\}$ as a basis for this vector space. Let us denote the subset of $V(f)$ consisting of those polynomials satisfying (7.11) as $R(f)$ ("the f-reducing polynomials"). Then $R(f)$ is in fact a subspace of $V(f)$, since

$$(s_1 h_1(x) + s_2 h_2(x))^q = s_1 h_1(x)^q + s_2 h_2(x)^q$$

by Lemmas 5.10 and 5.12. Now suppose $f(x)$ factors as follows:

$$f(x) = \prod_{i=1}^{m} p_i(x)^{e_i},$$

where the $p_i(x)$ are distinct irreducible monic polynomials.

Theorem 7.4. *The dimension of $R(f)$ is m, i.e., the number of distinct irreducible divisors of $f(x)$.*

Proof: As we have seen, a polynomial $h(x)$ is in $R(f)$ if and only if

$$f(x) \mid \prod_{s \in k} (h(x) - s).$$

Since the terms in this product are relatively prime, it follows that for each $i \in \{1, 2, \ldots, m\}$, there exists a unique $s_i \in k$ such that

$$h(x) \equiv s_i \pmod{p_i(x)^{e_i}}. \tag{7.16}$$

Conversely, given any collection $(s_1, s_2, \ldots, s_m)$ of elements of k, there exists a unique element in $V(f)$ satisfying (7.16). To see this, define for each $i = 1, 2, \ldots, m$,

$$f_i = p_i(x)^{e_i},$$

and

$$F_i = \prod_{\substack{j=1 \\ j \neq i}}^{m} f_j(x).$$

Clearly, $\gcd(F_i, f_i) = 1$, so there exists a polynomial G_i (unique, $\pmod{f_i}$) such that

$$F_i G_i \equiv 1 \pmod{f_i}.$$

Now define

$$h(x) = \sum_{i=1}^{m} s_i F_i G_i.$$

Then $h(x)$ satisfies (7.16).

Thus there is a one-to-one correspondence between the polynomials in $R(f)$ and the m–tuples $(s_1, s_2, \ldots, s_m)$. There are obviously q^m of the latter; and so also q^m of the former. ∎

Example 7.5. Let $f(x) = x^4 + x + 1$, $q = 2$. If $h(x) = h_0 + h_1 x + h_2 x^2 + h_3 x^3$, then the condition (7.11) is $h_0 + h_1 x^2 + h_2 x^4 + h_3 x^6 \equiv h_0 + h_1 x + h_2 x^2 + h_3 x^3$ $(\bmod\, x^4 + x + 1)$. But

$$x^4 \equiv x + 1 \pmod{x^4 + x + 1}$$

$$x^6 \equiv x^3 + x^2 \pmod{x^4 + x + 1},$$

and so (representing the polynomial $h_0 + h_1 + h_2 x^2 + h_3 x^3$ by the column vector $(h_0 h_1 h_2 h_3)^T$), this congruence will be satisfied if and only if

$$h_0 \begin{bmatrix} 1 \\ 0 \\ 0 \\ 0 \end{bmatrix} + h_1 \begin{bmatrix} 0 \\ 0 \\ 1 \\ 0 \end{bmatrix} + h_2 \begin{bmatrix} 1 \\ 1 \\ 0 \\ 0 \end{bmatrix} + h_3 \begin{bmatrix} 0 \\ 0 \\ 1 \\ 1 \end{bmatrix}$$

$$= h_0 \begin{bmatrix} 1 \\ 0 \\ 0 \\ 0 \end{bmatrix} + h_1 \begin{bmatrix} 0 \\ 1 \\ 0 \\ 0 \end{bmatrix} + h_2 \begin{bmatrix} 0 \\ 0 \\ 1 \\ 0 \end{bmatrix} + h_3 \begin{bmatrix} 0 \\ 0 \\ 0 \\ 1 \end{bmatrix}$$

i.e., if and only if $\mathbf{h} = [h_0 h_1 h_2 h_3]$ is in the nullspace of the matrix B, where

$$B = \begin{bmatrix} 1 & 0 & 1 & 0 \\ 0 & 0 & 1 & 0 \\ 0 & 1 & 0 & 1 \\ 0 & 0 & 0 & 1 \end{bmatrix} - \begin{bmatrix} 1 & 0 & 0 & 0 \\ 0 & 1 & 0 & 0 \\ 0 & 0 & 1 & 0 \\ 0 & 0 & 0 & 1 \end{bmatrix}$$

$$= \begin{bmatrix} 0 & 0 & 1 & 0 \\ 0 & 1 & 1 & 0 \\ 0 & 1 & 1 & 1 \\ 0 & 0 & 0 & 0 \end{bmatrix}.$$

An easy calculation puts B into row-reduced form:

$$B \sim \begin{bmatrix} 0 & 0 & 1 & 0 \\ 0 & 1 & 0 & 0 \\ 0 & 0 & 0 & 1 \\ 0 & 0 & 0 & 0 \end{bmatrix}.$$

Thus, in order to satisfy (7.11), we must have $h_2 = 0$, $h_1 = 0$, $h_3 = 0$. Hence the only solutions are $\mathbf{h} = [0000]$ and $[1000]$. Thus by Theorem 7.4, the dimension of $R(f)$ is one, and so $f(x)$ is the power of an irreducible polynomial. But since $f'(x) = 1$, $f(x)$ must be *irreducible* (See Problem 1d). ∎

Example 7.6. Now consider the factorization of $x^5 + x + 1$ over $GF(2)$. If $h(x) = h_0 + h_1 x + h_2 x^2 + h_3 x^3 + h_4 x^4$, then using the congruences

$$x^6 \equiv x^2 + x \pmod{x^5 + x + 1}$$
$$x^8 \equiv x^4 + x^3 \pmod{x^5 + x + 1},$$

we find that the conditions (7.11) become

$$h_0 \begin{bmatrix} 1 \\ 0 \\ 0 \\ 0 \\ 0 \end{bmatrix} + h_1 \begin{bmatrix} 0 \\ 0 \\ 1 \\ 0 \\ 0 \end{bmatrix} + h_2 \begin{bmatrix} 0 \\ 0 \\ 0 \\ 0 \\ 1 \end{bmatrix} + h_3 \begin{bmatrix} 0 \\ 1 \\ 1 \\ 0 \\ 0 \end{bmatrix} + h_4 \begin{bmatrix} 0 \\ 0 \\ 0 \\ 1 \\ 1 \end{bmatrix}$$

$$= h_0 \begin{bmatrix} 1 \\ 0 \\ 0 \\ 0 \\ 0 \end{bmatrix} + h_1 \begin{bmatrix} 0 \\ 1 \\ 0 \\ 0 \\ 0 \end{bmatrix} + h_2 \begin{bmatrix} 0 \\ 0 \\ 1 \\ 0 \\ 0 \end{bmatrix} + h_3 \begin{bmatrix} 0 \\ 0 \\ 0 \\ 1 \\ 0 \end{bmatrix} + h_4 \begin{bmatrix} 0 \\ 0 \\ 0 \\ 0 \\ 1 \end{bmatrix},$$

i.e., iff $\mathbf{h} = [h_0 h_1 h_2 h_3 h_4]$ is in the row-nullspace of the matrix B, where

$$B = \begin{bmatrix} 1 & 0 & 0 & 0 & 0 \\ 0 & 0 & 0 & 1 & 0 \\ 0 & 1 & 0 & 1 & 0 \\ 0 & 0 & 0 & 0 & 1 \\ 0 & 0 & 1 & 0 & 1 \end{bmatrix} - \begin{bmatrix} 1 & 0 & 0 & 0 & 0 \\ 0 & 1 & 0 & 0 & 0 \\ 0 & 0 & 1 & 0 & 0 \\ 0 & 0 & 0 & 1 & 0 \\ 0 & 0 & 0 & 0 & 1 \end{bmatrix}$$

$$= \begin{bmatrix} 0 & 0 & 0 & 0 & 0 \\ 0 & 1 & 0 & 1 & 0 \\ 0 & 1 & 1 & 1 & 0 \\ 0 & 0 & 0 & 1 & 1 \\ 0 & 0 & 1 & 0 & 0 \end{bmatrix}.$$

Thus the coefficients of $h(x)$ must satisfy the equations

$$\begin{aligned} h_1 \qquad + h_3 \qquad &= 0 \\ h_1 + h_2 + h_3 \qquad &= 0 \\ h_3 + h_4 &= 0 \\ h_2 \qquad\qquad &= 0 \end{aligned}$$

From these equations we conclude that $R(f)$ has dimension 2, and is spanned by the polynomials $[10000] = 1$ and $[01011] = x + x^3 + x^4$. Therefore the original polynomial has two distinct irreducible factors, and indeed it is easy to check that $\gcd(x^5+x+1, x^4+x^3+x) = x^3+x^2+1$, $\gcd(x^5+x+1, x^4+x^3+x+1) = x^2+x+1$. Hence $x^5+x+1 = (x^3+x^2+1)(x^2+x+1)$. We already know that these two polynomials are irreducible, and so this is the complete factorization. ∎

Berlekamp's algorithm can be considerably simplified if the polynomial to be factored is of the form $x^n - 1$ with $\gcd(n, q) = 1$.

Theorem 7.5. *A polynomial $h(x) = \sum_{i=0}^{n-1} h_i x^i$ satisfies congruence (7.11) (mod $x^n - 1$) if and only if $h_{iq} = h_i$ for all $i = 0, 1, \ldots, n-1$ (subscripts mod n).*

Proof: By Lemma 5.12,

$$h(x)^q \equiv \sum_{i=0}^{n-1} h_i x^{iq \bmod n} \pmod{x^n - 1}.$$

Comparing coefficients of x^{iq} on both sides of this congruence, we get the desired result. ∎

Now since $\gcd(q, n) = 1$, the mapping $i \to qi \pmod n$ is actually a *permutation* of the integers $\{0, 1, \ldots, n-1\}$. This permutation, like any permutation, can be written in cycle form. For example, if $n = 20$, $q = 3$, the permutation is:

$$\begin{bmatrix} 0 & 1 & 2 & 3 & 4 & 5 & 6 & 7 & 8 & 9 & 10 & 11 & 12 & 13 & 14 & 15 & 16 & 17 & 18 & 19 \\ 0 & 3 & 6 & 9 & 12 & 15 & 18 & 1 & 4 & 7 & 10 & 13 & 16 & 19 & 2 & 5 & 8 & 11 & 14 & 17 \end{bmatrix}$$

or in cycle form:

$$(0)(1,3,9,7)(2,6,18,14)(4,12,16,8)(5,15)(10)(11,13,19,17).$$

According to Theorem 7.5, therefore, any polynomial satisfying the congruence $h(x)^3 \equiv h(x) \pmod{x^{20} - 1}$ must be a $GF(3)$-linear combination of the following seven polynomials:

$$
\begin{aligned}
h_0(x) &= 1 \\
h_1(x) &= x + x^3 + x^7 + x^9 \\
h_2(x) &= x^2 + x^6 + x^{18} + x^{14} \\
h_4(x) &= x^4 + x^{12} + x^{16} + x^8 \\
h_5(x) &= x^5 + x^{15} \\
h_{10}(x) &= x^{10} \\
h_{11}(x) &= x^{11} + x^{13} + x^{17} + x^{19}
\end{aligned}
$$

The cycles of the permutation $i \to qi \pmod n$ are called *cyclotomic cosets.* The cyclotomic cosets also have another significance, which we now illustrate.

Since $3^4 \equiv 1 \pmod{20}$, it follows that $GF(81)$ contains an element α of order 20, and furthermore, that over $GF(81)$

$$x^{20} - 1 = \prod_{i=0}^{19}(x - \alpha^i).$$

As we have already seen (see Example 7.4), the factorization of $x^{20} - 1$ over $GF(3)$ can be determined from that of $x^{20} - 1$ over $GF(81)$ by computing minimal polynomials. For example, the minimal polynomial of α over $GF(3)$ is

$$f_1(x) = (x - \alpha)(x - \alpha^3)(x - \alpha^9)(x - \alpha^7).$$

The exponents of α, viz. $1, 3, 9, 7$ are just the elements of the cyclotomic coset $(1, 3, 9, 7)$ computed above! Also notice that each element of $(1, 3, 7, 9)$ is relatively prime to 20, and so by Lemma 5.4, each $\alpha^i, i = 1, 3, 7, 9$ will have order 20. Hence $f_1(x) = (x - \alpha)(x - \alpha^3)(x - \alpha^7)(x - \alpha^9)$ is an irreducible divisor not only of $x^{20} - 1$ but in fact of the cyclotomic polynomial $\Phi_{20}(x)$. Similarly, if we denote by $f_i(x)$ the minimal polynomial of α^i, we see that $f_i(x)$ is an irreducible divisor of $\Phi_{n/\gcd(n,i)}(x)$. If we now denote the cyclotomic coset containing i by C_i, we get the following table:

i	C_i	$\mid C_i \mid = \deg f_i(x)$	$20/\gcd(20,i)$
0	(0)	1	1
1	$(1,3,9,7)$	4	20
2	$(2,6,18,14)$	4	10
4	$(4,12,16,8)$	4	5
5	$(5,15)$	2	4
10	(10)	1	2
11	$(11,13,19,17)$	4	20

From this table we can draw the following conclusions about the factors of $x^{20} - 1$, which we can state in terms of the cyclotomic polynomials $\Phi_d(x)$, $d \mid 20$:

d	$\Phi_d(x)$	Factorization
1	$x - 1$	$f_1(x)$ (irreducible)
2	$x + 1$	$f_{10}(x)$ (irreducible)
4	$x^2 + 1$	$f_5(x)$ (irreducible)
5	$x^4 + x^3 + x^2 + x + 1$	$f_4(x)$ (irreducible)
10	$x^4 - x^3 + x^2 - x + 1$	$f_2(x)$ (irreducible)
20	$x^8 - x^6 + x^4 - x^2 + 1$	$f_1(x) \cdot f_{11}(x)$

This gives the complete factorization, except for $\Phi_{20}(x)$. To factor $\Phi_{20}(x)$ explicitly, we will use Berlekamp's algorithm.

According to Theorem 7.3, we have, for any of the polynomials $h_i(x)$ listed above, that

$$(x^{20} - 1) = \gcd(x^{20} - 1, h_i(x)) \cdot \gcd(x^{20} - 1, h_i(x) + 1) \cdot \gcd(x^{20} - 1, h_i(x) + 2).$$

But since $\Phi_{20}(x) \mid (x^{20} - 1)$, it follows that $h_i^q \equiv h_i \pmod{\Phi_{20}(x)}$ as well, and hence

$$\Phi_{20}(x) = \gcd(\Phi_{20}, h_i) \cdot \gcd(\Phi_{20}, h_i + 1) \cdot \gcd(\Phi_{20}, h_i + 2).$$

Let us now compute these gcd's with $i = 1$, using Euclid's algorithm.

First we compute $\gcd(\Phi_{20}, h_1) = \gcd(x^8 - x^6 + x^4 - x^2 + 1, x^9 + x^7 + x^3 + x)$ using "synthetic division" (i.e., keeping track of the coefficients without writing down the powers of x). Also we write $\Phi_{20}(x) = x^8 + 2x^6 + x^4 + 2x^2 + 1$ to remind ourselves that the calculation is done in $GF(3)$. First we divide $h_1(x)$ by $\Phi_{20}(x)$:

$$
\begin{array}{r|l}
 & 1\ 0 \\
\cline{2-2}
1\ 0\ 2\ 0\ 1\ 0\ 2\ 0\ 1 & 1\ 0\ 1\ 0\ 0\ 0\ 1\ 0\ 1\ 0 \\
 & 1\ 0\ 2\ 0\ 1\ 0\ 2\ 0\ 1\ 0 \\
\cline{2-2}
 & 2\ 0\ 2\ 0\ 2\ 0\ 0\ 0
\end{array}
$$

Hence $\gcd(h_1(x), \Phi_{20}(x)) = \gcd(\Phi_{20}(x), 2x^6 + 2x^4 + 2x^2) = \gcd(\Phi_{20}, x^4 + x^2 + 1)$. Now $x^4 + x^2 + 1$ has degree 4, which is the degree of one of the irreducible factors of $\Phi_{20}(x)$. But $x^4 + x^2 + 1$ isn't irreducible (it has $x - 1$ as a factor), so we can stop. We know without further calculation that

$$\gcd(\Phi_{20}, h_1) = 1.$$

Next we consider the calculation of $\gcd(\Phi_{20}, h_1 + 1)$. Letting $r_{-1}(x) = h_1(x) + 1$, $r_0(x) = \Phi_{20}(x)$, and letting r_{i+1} be the remainder obtained when r_{i-1} is divided by r_i, we obtain the following sequence:

$$
\begin{aligned}
r_{-1} &= 1010001011 \\
r_0 &= 102010201 \\
r_1 &= 20202001 \\
r_2 &= 1000211 \\
r_3 &= 102222 \\
r_4 &= 11021 \\
r_5 &= 0
\end{aligned}
$$

Hence $\gcd(\Phi_{20}, h_1 + 1) = 11021 = x^4 + x^3 + 2x + 1$, and this must be one of the irreducible divisors of Φ_{20}. To compute the other, we could of course compute $\gcd(\Phi_{20}, h_1 + 2)$; or we could divide Φ_{20} by $x^4 + x^3 + 2x + 1$. However, we can avoid *any* computation, by noticing that in the factorization $\Phi_{20}(x) =$

$f_1(x) \cdot f_{11}(x)$, the roots of $f_{11}(x)$ (i.e., $\alpha^{11}, \alpha^{13}, \alpha^{19}, \alpha^{17}$) are just the *inverses* of the roots of $f_1(x)$ (i.e., $\alpha, \alpha^3, \alpha^9, \alpha^7$). There is a simple general relationship between polynomials whose roots are the reciprocals of each other. Let

$$f(x) = a_0 + a_1 x + \cdots + a_m x^m,$$

and

$$\tilde{f}(x) = x^m f(1/x) = a_0 x^m + a_1 x^{m-1} + \cdots + a_m. \tag{7.17}$$

Then clearly $f(\alpha) = 0$ if and only if $\tilde{f}(\alpha^{-1}) = 0$. Thus a polynomial with roots which are the *reciprocals* of those of $f(x)$ can be obtained merely by writing the coefficients of $f(x)$ in the reverse order! For the present case if $f(x) = 11021$, then $\tilde{f}(x) = 12011$, and so the complete factorization of $\Phi_{20}(x)$ over $GF(3)$ is

$$\Phi_{20}(x) = (x^4 + x^3 + 2x + 1)(x^4 + 2x^3 + x + 1).$$

Let us now take one last look at the irreducible polynomials of degree 6 over $GF(2)$. At the beginning of the chapter, we saw that

$$V_6(x) = \Phi_9(x)\Phi_{21}(x)\Phi_{63}(x),$$

and concluded that $\Phi_9(x) = x^6 + x^3 + 1$ is irreducible. However, the other two polynomials are reducible. Now an easy calculation gives

$$\Phi_{21}(x) = x^{12} + x^{11} + x^9 + x^8 + x^6 + x^4 + x^3 + x + 1,$$

which must factor over $GF(2)$ into the product of two degree 6 irreducibles. With $h(x) = x^3 + x^6 + x^{12}$ (since (3,6,12) is a cyclotomic coset (mod 21)), a simple calculation gives $\gcd(\Phi_{21}, h) = x^6 + x^5 + x^4 + x^2 + 1$, and so by our remarks above, the other irreducible factor of Φ_{21} must be $x^6 + x^4 + x^2 + x + 1$. Hence

$$\Phi_{21}(x) = (x^6 + x^5 + x^4 + x^2 + 1) \cdot (x^6 + x^4 + x^2 + x + 1).$$

On to $\Phi_{63}(x)$! By Theorem 7.1a, $\Phi_{63}(x) = \Phi_{21}(x^3)$, and so from the above factorization of $\Phi_{21}(x)$ we get immediately

$$\Phi_{63}(x) = (x^{18} + x^{15} + x^{12} + x^6 + 1) \cdot (x^{18} + x^{12} + x^6 + x^3 + 1).$$

Each of these 18th degree polynomials factors into three irreducibles of degree 6. We leave the rest of the factorization to the reader.

Problems for Chapter 7.

1. Prove that if n is odd, $\Phi_{2n}(x) = \Phi_n(-x)$.

2. Notice that for all of the cyclotomic polynomials we computed in the text, except $\Phi_1(x)$, the constant term is $+1$. Is this a coincidence, or a theorem?

3. Show that the cyclotomic polynomials themselves satisfy (7.17).

4. Calculate the following cyclotomic polynomials.

a. $\Phi_{24}(x)$.
b. $\Phi_{35}(x)$.
c. $\Phi_{40}(x)$.
d. $\Phi_{60}(x)$.
e. $\Phi_{105}(x)$.

5. Factor the following cyclotomic polynomials explicitly, over the given finite field.

a. $\Phi_{17}(x)$, over $GF(2)$.
b. $\Phi_{11}(x)$, over $GF(3)$.
c. $\Phi_{13}(x)$, over $GF(5)$.
d. $\Phi_{19}(x)$, over $GF(7)$.

6. Let F be a field of characteristic 2. Calculate
  a. $\Phi_{10}(x)$.
  b. $\Phi_{12}(x)$.
  c. $\Phi_{14}(x)$.
  d. $\Phi_{16}(x)$.

7. Factor $x^{16} - x$ into irreducible factors:
  a. Over $GF(16)$.
  b. Over $GF(2)$.
  c. Over $GF(4)$. (Represent $GF(4)$ as $\{0, 1, \omega, \omega^2\}$, where $\omega^2 = \omega + 1$.)

8. Find the complete factorization of $x^{24} - 1$ over:
  a. $GF(2)$.
  b. $GF(3)$.
  c. $GF(4)$.
  d. $GF(5)$.
  e. $GF(7)$.

9. Determine the shape of the factorization of $\Phi_{360}(x)$ over $GF(3)$. (Cf. Example 7.4.)

10. In Example 6.1 we claimed that "the techniques to be developed in Chapter 7" would allow us to produce the complete factorization of $x^{16} + x$ over $GF(2)$. Use the techniques of this chapter to do this.

11. In the text we obtained the following partial factorization of $\Phi_{63}(x)$ over $GF(2)$:

$$\Phi_{63}(x) = (x^{18} + x^{15} + x^{12} + x^6 + 1) \cdot (x^{18} + x^{12} + x^6 + x^3 + 1),$$

and predicted that both of these 18th degree polynomials would factor into three irreducibles of degree 6. Find these six degree 6 irreducible factors of Φ_{63} explicitly, using Berlekamp's algorithm.

Chapter 8

Trace, Norm, and Bit-Serial Multiplication

In this chapter we will study *traces* and *norms*, which are very useful analytic tools in finite fields. Traces and norms can be computed in any field, finite or not; for example in the field of complex numbers the trace of z equals twice the real part of z, and the norm of z equals the square of the absolute value of z. In finite fields, however, the definition is subtler and the results more useful.

Let $F = GF(q)$, $K = GF(q^n)$. Then by Theorem 6.6, we may view F as a subfield of K. If α is an element of K, its *trace relative to the subfield F* is defined as follows:

$$\text{Tr}_F^K(\alpha) = \alpha + \alpha^q + \alpha^{q^2} + \cdots + \alpha^{q^{n-1}}. \tag{8.1}$$

The *norm* is defined as follows:

$$\text{N}_K^F(\alpha) = \alpha \cdot \alpha^q \cdots \alpha^{q^{n-1}}. \tag{8.2}$$

We will sometimes talk about traces and norms relative to different subfields of K simultaneously (e.g. Theorem 8.2, below), and in such situations it will be important to keep the sub- and superscripts on the Tr's and N's. But when no confusion is likely to arise, we will omit them, as in the following theorem.

Theorem 8.1. *For all $\alpha, \beta \in K$ we have*

(*a*) $\mathrm{Tr}(\alpha) \in F$	(*a′*) $\mathrm{N}(\alpha) \in F$
(*b*) $\mathrm{Tr}(\alpha+\beta) = \mathrm{Tr}(\alpha) + \mathrm{Tr}(\beta)$	(*b′*) $\mathrm{N}(\alpha\beta) = \mathrm{N}(\alpha)\,\mathrm{N}(\beta)$
(*c*) $\mathrm{Tr}(\lambda\alpha) = \lambda\,\mathrm{Tr}(\alpha)$ *if* $\lambda \in F$	(*c′*) $\mathrm{N}(\lambda\alpha) = \lambda^n\,\mathrm{N}(\alpha)$ *if* $\lambda \in F$
(*d*) $\mathrm{Tr}(\alpha^q) = \mathrm{Tr}(\alpha)$	(*d′*) $\mathrm{N}(\alpha^q) = \mathrm{N}(\alpha)$
(*e*) Tr *maps* K *onto* F	(*e′*) N *maps* K *onto* F

Proof: We prove only the results about the trace; the proofs for the norm are similar, and are left as an exercise.
(a) Since $\alpha^{q^n} = \alpha$, we have $\mathrm{Tr}(\alpha)^q = (\alpha + \alpha^q + \cdots + \alpha^{q^{n-1}})^q = \alpha^q + \alpha^{q^2} + \cdots + \alpha^{q^{n-1}} + \alpha = \mathrm{Tr}(\alpha)$. Thus by Lemma 5.10, $\mathrm{Tr}(\alpha) \in F$.
(b) Follows from the fact that $(\alpha+\beta)^{q^i} = \alpha^{q^i} + \beta^{q^i}$ (Lemma 5.12).
(c) Follows from the fact that if $\lambda \in F$, then $\lambda^{q^i} = \lambda$ (Lemma 5.10).
(d) Since $\alpha^{q^n} = \alpha$ (Lemma 5.10), by (8.1) we have

$$\begin{aligned}\mathrm{Tr}(\alpha^q) &= \alpha^q + \alpha^{q^2} + \cdots + \alpha^{q^{n-1}} + \alpha^{q^n} \\ &= \alpha^q + \alpha^{q^2} + \cdots + \alpha^{q^{n-1}} + \alpha \\ &= \mathrm{Tr}(\alpha)\end{aligned}$$

(e) What this means is that given $\lambda \in F$, there exists $\alpha \in K$ such that $\mathrm{Tr}(\alpha) = \lambda$. We can prove even more by noting that since by (a), $\mathrm{Tr}(\alpha) \in F$, we have the factorization

$$x^{q^n} - x = \prod_{\lambda \in F} (x + x^q + \cdots + x^{q^{n-1}} - \lambda), \tag{8.3}$$

since every root of the left side is a root of exactly one of the factors on the right. Since however $x^{q^n} - x = \prod(x - \alpha : \alpha \in K)$ (Theorem 6.1), it follows that each of the q polynomials $x + x^q + \cdots + x^{q^{n-1}} - \lambda$ has exactly q^{n-1} roots in K. This proves (e). ∎

Example 8.1. Let $K = GF(2^4)$, $F = GF(2)$. We define K via the irreducible polynomial $f(x) = x^4 + x^3 + x^2 + x + 1 = \Phi_5(x)$. Letting α denote the element $x \bmod f(x)$, (alternatively α is a root of the equation $f(x) = 0$

in K) we know that every element $\beta \in K$ can be expressed uniquely as $\beta = \beta_0 + \beta_1\alpha + \beta_2\alpha^2 + \beta_3\alpha^3$. Then by parts (b) and (c) of Theorem 8.1, we have

$$\mathrm{Tr}(\beta) = \beta_0 \,\mathrm{Tr}(1) + \beta_1 \,\mathrm{Tr}(\alpha) + \beta_2 \,\mathrm{Tr}(\alpha^2) + \beta_3 \,\mathrm{Tr}(\alpha^3).$$

Thus in order to compute $\mathrm{Tr}(\beta)$ for any β it is sufficient to compute $\mathrm{Tr}(1)$, $\mathrm{Tr}(\alpha)$, $\mathrm{Tr}(\alpha^2)$, and $\mathrm{Tr}(\alpha^3)$. Clearly $\mathrm{Tr}(1) = 1+1+1+1 = 0$. Now $\mathrm{Tr}(\alpha) = \alpha + \alpha^2 + \alpha^4 + \alpha^8$; this could be computed in a straightforward manner in $GF(16)$. But notice that if $f(x) = (x-\alpha)(x-\alpha^2)(x-\alpha^4)(x-\alpha^8) = x^4 + c_1x^3 + c_2x^2 + c_3x + c_4$ is the minimal polynomial of α, then $\mathrm{Tr}(\alpha) = c_1$, and by construction $f(x) = x^4 + x^3 + x^2 + x + 1$, and so $\mathrm{Tr}(\alpha) = 1$. Next we have $\mathrm{Tr}(\alpha^2) = \mathrm{Tr}(\alpha) = 1$, by part (d) of Theorem 8.1. Finally note that since $f(x) \mid \Phi_5(x)$, $\alpha^5 = 1$, and so $\alpha^3 = \alpha^8$, i.e., α^3 is conjugate to α, and so $\mathrm{Tr}(\alpha^3) = \mathrm{Tr}(\alpha) = 1$. Summarizing, we have shown without any calculation that

$$\mathrm{Tr}(1) = 0,\ \mathrm{Tr}(\alpha) = \mathrm{Tr}(\alpha^2) = \mathrm{Tr}(\alpha^3) = 1,$$

and so, if $\beta = (\beta_0, \beta_1, \beta_2, \beta_3) = \beta_0 + \beta_1\alpha + \beta_2\alpha^2 + \beta_3\alpha^3$,

$$\begin{aligned} \mathrm{Tr}(\beta) &= \beta_1 + \beta_2 + \beta_3 \\ &= \beta \cdot (0111) \qquad \text{(dot product)}. \end{aligned}$$

Similarly, if we had instead chosen to build $GF(16)$ with the irreducible polynomial $f(x) = x^4 + x + 1$, we would compute

$$\mathrm{Tr}(1) = 1+1+1+1 = 0$$

$$\mathrm{Tr}(\alpha) = (\text{coefficient of } x^3 \text{ in } f(x)) = 0$$

$$\mathrm{Tr}(\alpha^2) = \mathrm{Tr}(\alpha) = 0 \quad (\alpha^2 \text{ is conjugate to } \alpha).$$

This leaves only $\mathrm{Tr}(\alpha^3)$. But if $\mathrm{Tr}(\alpha^3) = 0$, it would follow that $\mathrm{Tr}(\beta) = 0$ identically, and this contradicts part (d) of Theorem 8.1. Hence, again without any calculation, we find that relative to this coordinate system, in $GF(16)$ we have

$$\begin{aligned}\mathrm{Tr}(\beta) &= \beta_3 \\ &= \beta \cdot (0001).\end{aligned}$$ ∎

In Theorem 8.1 there were only two fields, F and K. We come now to a result about how the trace and norm behave when *three* fields are present. Let $F \subseteq E \subseteq K$ be three finite fields, say

$$\begin{aligned} K &\simeq GF(q^n) \\ E &\simeq GF(q^d) \quad (d \text{ a divisor of } n) \\ F &\simeq GF(q). \end{aligned}$$

In this situation there are three traces to consider:

$$\begin{aligned} \text{The } \textit{great trace } \mathrm{Tr}_F^K(\alpha) &= \alpha + \alpha^q + \cdots + \alpha^{q^{n-1}} \\ \text{The } \textit{lesser trace } \mathrm{Tr}_F^E(\alpha) &= \alpha + \alpha^q + \cdots + \alpha^{q^{d-1}} \\ \text{The } \textit{relative trace } \mathrm{Tr}_E^K(\alpha) &= \alpha + \alpha^{q^d} + \cdots + \alpha^{q^{d((n/d)-1)}}. \end{aligned}$$

Also, there are three norms:

$$\begin{aligned} \text{The } \textit{great norm } \mathrm{N}_F^K(\alpha) &= \alpha^{1+q+\cdots+q^{n-1}} \\ \text{The } \textit{lesser norm } \mathrm{N}_F^E(\alpha) &= \alpha^{1+q+\cdots+q^{d-1}} \\ \text{The } \textit{relative norm } \mathrm{N}_E^K(\alpha) &= \alpha^{1+q^d+\cdots+q^{d((n/d)-1)}}. \end{aligned}$$

The next theorem gives the relationship between all these norms and traces.

Theorem 8.2.

$$\begin{aligned} \mathrm{Tr}_F^K(\alpha) &= \mathrm{Tr}_F^E(\mathrm{Tr}_E^K(\alpha)), \\ \mathrm{N}_F^K(\alpha) &= \mathrm{N}_F^E(\mathrm{N}_E^K(\alpha)). \end{aligned}$$

Proof: (Sketch only.) We illustrate the proof for the trace with $n = 6, d = 3$. We leave both the formal proofs to the reader. When $n = 6, d = 3$, we have, from the definitions,

$$\begin{aligned}
\mathrm{Tr}_F^K(\alpha) &= \alpha + \alpha^q + \alpha^{q^2} + \alpha^{q^3} + \alpha^{q^4} + \alpha^{q^5} \\
\mathrm{Tr}_F^E(\alpha) &= \alpha + \alpha^q + \alpha^{q^2} \\
\mathrm{Tr}_E^K(\alpha) &= \alpha + \alpha^{q^3}.
\end{aligned}$$

Thus

$$\begin{aligned}
\mathrm{Tr}_F^E(\mathrm{Tr}_E^K(\alpha)) &= (\alpha + \alpha^{q^3}) + (\alpha + \alpha^{q^3})^q + (\alpha + \alpha^{q^3})^{q^2} \\
&= \alpha + \alpha^{q^3} + \alpha^q + \alpha^{q^4} + \alpha^{q^2} + \alpha^{q^5} \\
&= \mathrm{Tr}_F^K(\alpha).
\end{aligned}$$ ∎

We continue to assume that $F \subseteq K$ are finite fields with $F \cong GF(q)$, $K \cong GF(q^n)$, and let $\alpha \in K$. We saw in Example 8.1 that in one particular case the trace of α could be determined from the coefficients of the minimal polynomial of α. In the next theorem, we will see that $\mathrm{Tr}_F^K(\alpha)$ and $\mathrm{N}_F^K(\alpha)$ can always be expressed in terms of the coefficients of the minimal polynomials of α. Thus let

$$\alpha, \alpha^q, \alpha^{q^2}, \ldots, \alpha^{q^{d-1}}, \quad (d \mid n)$$

be the conjugates of α (with respect to the subfield F). Then the minimal polynomial of α is, by definition,

$$(8.4) \qquad (x - \alpha)(x - \alpha^q) \ldots (x - \alpha^{q^{d-1}}) = x^d + c_1 x^{d-1} + \cdots + c_d,$$

where the coefficients $c_1, c_2, \ldots, c_d$ can in principle be determined by multiplying out the d factors on the left.

Theorem 8.3.

$$\mathrm{Tr}_F^K(\alpha) = -\tfrac{n}{d}c_1$$

$$\mathrm{N}_F^K(\alpha) = c_d^{n/d}.$$

Proof: Since $d \mid n$, we know by Theorem 6.6 that $E = GF(q^d)$ is a subfield of $K = GF(q^n)$. Furthermore, $\alpha \in E$, since $\alpha^{q^d} = \alpha$. It follows from Theorem 8.2 that

$$\mathrm{Tr}_F^K(\alpha) = \mathrm{Tr}_F^E(\mathrm{Tr}_E^K(\alpha)).$$

But since $\alpha \in E$, $\mathrm{Tr}_E^K(\alpha) = (n/d) \cdot \alpha$. (This follows from the general result, which we leave as an exercise, that if $E = GF(q)$, $K = GF(q^m)$, and if $\alpha \in E$, then $\mathrm{Tr}_E^K(\alpha) = m \cdot \alpha$.) Thus by Theorem 8.1(c),

$$\begin{aligned}\mathrm{Tr}_F^K(\alpha) &= \mathrm{Tr}_F^E\ \tfrac{n}{d}\alpha \\ &= \tfrac{n}{d}\,\mathrm{Tr}_F^E(\alpha).\end{aligned}$$

But from Eq. (8.4) we have

$$c_1 = -\alpha - \alpha^q \cdots - \alpha^{q^{d-1}} = -\,\mathrm{Tr}_F^E(\alpha).$$

Hence $\mathrm{Tr}_F^K(\alpha) = -(n/d) \cdot c_1$, as claimed. The proof for $\mathrm{N}_F^K(\alpha)$ is similar and is left as an exercise. ∎

Example 8.2. (Cf. Example 5.6.) Let $K = GF(2^4), F = GF(2)$, with defining polynomial $f(x) = x^4 + x + 1$. Using Theorem 8.3, we construct the following table:

i	min. poly of α^i	$\mathrm{Tr}(\alpha^i)$	Comments
0	$x+1$	0	$c_1 = 1, n/d = 4.$
1	x^4+x+1	0	$c_1 = 0, n/d = 1.$
2	x^4+x+1	0	α^2 conjugate to α.
3	$x^4+x^3+x^2+x+1$	1	$c_1 = 1, n/d = 1.$
4	x^4+x+1	0	conjugate to α.
5	x^2+x+1	0	$c_1 = 1, n/d = 2.$
6	$x^4+x^3+x^2+x+1$	1	conjugate to α^3.
7	x^4+x^3+1	1	$c_1 = 1, n/d = 1.$
8	x^4+x+1	0	conjugate to α.
9	$x^4+x^3+x^2+x+1$	1	conjugate to α^3.
10	x^2+x+1	0	conjugate to α^5.
11	x^4+x^3+1	1	conjugate to α^7.
12	$x^4+x^3+x^2+x+1$	1	conjugate to α^3.
13	x^4+x^3+1	1	conjugate to α^7.
14	x^4+x^3+1	1	conjugate to α^7.

Most of the "Comments" are self-explanatory. For example, the $i = 11$ entry says that α^{11}, being conjugate to α^7 (since $(\alpha^7)^8 = \alpha^{11}$) must by Theorem 8.1d have the same trace as α^7; and $\mathrm{Tr}(\alpha^7) = 1$ was previously determined. Similarly, the $i = 5$ entry points out that since the minimal polynomial of α^5 is $x^2 + x + 1$ we can apply Theorem 8.3 with $n = 4, d = 2$, and $c_1 = 1$ to get $\mathrm{Tr}(\alpha^5) = 2 \cdot 1 = 0$. ∎

Our next results characterize those elements $\alpha \in K$ with $\mathrm{Tr}(\alpha) = 0$, (also $\mathrm{N}(\alpha) = 1$).

Theorem 8.4. *(As before $K = GF(q^n)$, $F = GF(q)$.)* $\mathrm{Tr}(\alpha) = 0$ *if and only if there exists an element* $\beta \in K$ *such that*

$$\alpha = \beta - \beta^q.$$

Similarly $\mathrm{N}(\alpha) = 1$ *if and only if there exists an element* $\beta \in K^*$ *such that*

$$\alpha = \beta^{1-q}.$$

Proof: (As usual, we give proofs for Trace only, leaving the proofs for Norm left as a problem.)

First notice that if α is of the form $\beta - \beta^q$, then by Theorem 8.1d $\mathrm{Tr}(\alpha) = \mathrm{Tr}(\beta - \beta^q) = \mathrm{Tr}(\beta) - \mathrm{Tr}(\beta^q) = \mathrm{Tr}(\beta) - \mathrm{Tr}(\beta) = 0$. Thus every element of this form has trace zero. The question to be resolved is whether any elements not of this form can have trace 0. Now the mapping $L : \beta \to \beta - \beta^q$ is a linear transformation of the vector space K into itself. Its kernel is the set $L_0 = \{\beta : \beta = \beta^q\}$, i.e., the subfield F, which is a one-dimensional subspace of K. It follows from linear algebra that the image set of L is an $n - 1$ dimensional subspace of K, i.e., there are q^{n-1} elements of the form $\beta - \beta^q$. But we saw in the proof of Theorem 8.1 that there are exactly q^{n-1} elements in K of trace 0. It follows that they all must be of the form $\beta - \beta^q$. ∎

The proof of Theorem 8.4 we have given is short, but not constructive, in the sense that it gives no efficient procedure for "solving" the equation $\alpha = \beta - \beta^q$ for β. Hilbert gave a constructive proof, based on the following lemma.

Lemma 8.5. *Let* α, θ *be elements of* K. *If* β *is defined by*

$$\beta = \alpha\theta^q + (\alpha + \alpha^q)\theta^{q^2} + \cdots + (\alpha + \alpha^q + \cdots + \alpha^{q^{n-2}})\theta^{q^{n-1}}, \tag{8.5}$$

then

$$\beta - \beta^q = \alpha(\mathrm{Tr}(\theta) - \theta) - \theta(\mathrm{Tr}(\alpha) - \alpha).$$

Proof: By definition of β,

$$\beta^q = \alpha^q\theta^{q^2} + (\alpha^q + \alpha^{q^2})\theta^{q^3} + \cdots + (\alpha^q + \alpha^{q^2} + \cdots + \alpha^{q^{n-1}})\theta^{q^n}.$$

Thus by subtraction,

$$\beta - \beta^q = \alpha(\theta^q + \theta^{q^2} + \cdots + \theta^{q^{n-1}}) - (\alpha^q + \alpha^{q^2} + \cdots + \alpha^{q^{n-1}})\theta^{q^n}.$$

But $\theta^{q^n} = \theta$, and

$$\theta^q + \cdots + \theta^{q^{n-1}} = \mathrm{Tr}(\theta) - \theta.$$

Similarly,

$$\alpha^q + \alpha^{q^2} + \cdots + \alpha^{q^{n-1}} = \mathrm{Tr}(\alpha) - \alpha. \quad \blacksquare$$

Now we can give Hilbert's constructive proof of Theorem 8.4. Choose θ to be a fixed element of K with $\mathrm{Tr}(\theta) = 1$ (such a θ exists, because of Theorem 8.1(e)). Then if $\mathrm{Tr}(\alpha) = 0$, the element β defined by (8.5) satisfies $\alpha = \beta - \beta^q$, by Lemma 8.5.

Our main interest in Theorem 8.4 is that it tells us how to solve the quadratic equations over fields of characteristic 2. (This is important, since the ordinary quadratic formula for solving the equation $Ax^2 + Bx + C = 0$, viz. $x = (-B \pm \sqrt{B^2 - 4AC})/2A$ is useless in characteristic 2!) Indeed, when the small field F is $GF(2)$, and the big field K is $GF(2^m)$, Theorem 8.4 says that if $\alpha \in GF(2^m)$, the equation

$$x^2 + x = \alpha \tag{8.6}$$

has a solution if and only if $\mathrm{Tr}(\alpha) = 0$. Thus if $\mathrm{Tr}(\alpha) = 1$, there are no solutions to (8.6). If however $\mathrm{Tr}(\alpha) = 0$, Theorem 8.4 guarantees that there are solutions. In fact, there are exactly two solutions. One solution is given by $x = \beta$, where β is given by (8.5) for a fixed element $\theta \in GF(2^m)$ with $\mathrm{Tr}(\theta) = 1$. Also notice that if $x = \beta$ is one solution, then $x = \beta + 1$ is another, since $(\beta + 1)^2 + (\beta + 1) = \beta^2 + \beta = \alpha$. Furthermore, despite appearances, the mapping $\alpha \to \beta$ given by (8.5) is very simple to implement, because it is a *linear* mapping. We illustrate this in the following example.

Example 8.3. Consider the field $GF(16)$, which contains a primitive root α

satisfying $\alpha^4 + \alpha + 1 = 0$. The following table summarizes the field structure.

i	α^i	
0	0001	
1	0010	
2	0100	
3	1000	
4	0011	
5	0110	
6	1100	
7	1011	
8	0101	
9	1010	$\leftarrow$ *e.g.*, $\alpha^9 = \alpha^3 + \alpha$
10	0111	
11	1110	
12	1111	
13	1101	
14	1001	

We now consider the equation

$$y^2 + y = \gamma, \tag{8.7}$$

where γ is a given element of $GF(16)$. According to Theorem 8.4, this equation will have a solution $y \in GF(16)$ if and only if $\mathrm{Tr}(\gamma) = 0$. But if

$$\gamma = \gamma_3\alpha^3 + \gamma_2\alpha^2 + \gamma_1\alpha + \gamma_0 = (\gamma_3\gamma_2\gamma_1\gamma_0),$$

$\gamma_i \in GF(2)$, then since (see Example 8.1), $\mathrm{Tr}(1) = \mathrm{Tr}(\alpha) = \mathrm{Tr}(\alpha^2) = 0$, $\mathrm{Tr}(\alpha^3) = 1$, it follows that

$$\mathrm{Tr}(\gamma) = \gamma_3. \tag{8.8}$$

Thus (8.7) has a solution iff $\gamma_3 = 0$, i.e., iff $\gamma = 0$ or $\gamma = \alpha^i$ with $i = 0, 1, 2, 4, 5, 8, 10$. Let us now use Hilbert's construction (8.5) to solve the same equation. First we need an element θ with trace 1. From the table, we see that $\mathrm{Tr}(\alpha^3) = 1$, and so we can choose $\theta = \alpha^3$. Now (8.5) can be written as

$$\begin{aligned} y &= \gamma\theta^2 + (\gamma + \gamma^2)\theta^4 + (\gamma + \gamma^2 + \gamma^4)\theta^8 \\ &= \gamma(\theta^2 + \theta^4 + \theta^8) + \gamma^2(\theta^4 + \theta^8) + \gamma^4\theta^8. \end{aligned}$$

With $\theta = \alpha^3$, one easily verifies that

$$\theta^2 + \theta^4 + \theta^8 = \alpha^3 + 1 = (1001)$$

$$\theta^4 + \theta^8 = \alpha^2 + 1 = (0101)$$

$$\theta^8 = \alpha^3 + \alpha = (1010).$$

Thus a solution to (8.6) when $\mathrm{Tr}(\gamma) = 0$ is

$$y = \gamma(\alpha^3 + 1) + \gamma^2(\alpha^2 + 1) + \gamma^4(\alpha^3 + \alpha). \tag{8.9}$$

Now as mentioned above the mapping $\gamma \to y$ of (8.9) is linear, so to describe its behavior in general it is sufficient to describe its behavior on the basis elements $1, \alpha, \alpha^2, \alpha^3$. A simple computation gives (use (8.9)):

$$\gamma = 1 : y = \alpha^2 + \alpha = (0110)$$

$$\gamma = \alpha : y = \alpha^3 + \alpha + 1 = (1011)$$

$$\gamma = \alpha^2 : y = \alpha^3 + 1 = (1001)$$

$$\gamma = \alpha^3 : y = 1 = (0001).$$

Thus if $\gamma = (\gamma_3\gamma_2\gamma_1\gamma_0)$, a solution to (8.7) is

$$y = (\gamma_3\gamma_2\gamma_1\gamma_0) \cdot \begin{bmatrix} 0 & 0 & 0 & 1 \\ 1 & 0 & 0 & 1 \\ 1 & 0 & 1 & 1 \\ 0 & 1 & 1 & 0 \end{bmatrix} \tag{8.10}$$

For example, to solve $y^2 + y = \alpha^{10} = (0111)$, we use (8.10) and obtain

$$y = (0\,1\,1\,1)\begin{bmatrix} 0 & 0 & 0 & 1 \\ 1 & 0 & 0 & 1 \\ 1 & 0 & 1 & 1 \\ 0 & 1 & 1 & 0 \end{bmatrix} = (0\,1\,0\,0) = \alpha^2.$$

As a check, $y^2 + y = \alpha^4 + \alpha^2 = \alpha^2 + \alpha + 1 = \alpha^{10}$. The other solution to $y^2 + y = \alpha^{10}$ is naturally $\alpha^2 + 1 = (0101)$. ∎

Example 8.4. Now consider solving Eq. (8.6) in the field $GF(2^5)$. Here we may simply take $\theta = 1$, and Hilbert's solution boils down to

$$y = \alpha^2 + \alpha^8.$$

More generally in $GF(2^n)$ with *odd* n, a solution to (8.6) is given by

$$y = \alpha^2 + \alpha^8 + \cdots + \alpha^{2^{n-2}}.$$ ∎

Having now discussed the special quadratic equation (8.7), it is not much more difficult to consider the general quadratic equation in a finite field of characteristic 2:

$$Ax^2 + Bx + C = 0, \tag{8.11}$$

with $A, B, C \in GF(2^n)$. Naturally we assume $A \neq 0$; otherwise the equation isn't quadratic!

• Case 1. $B \neq 0$. Here if we set $y = AB^{-1} \cdot x$, $\gamma = ACB^{-2}$, (8.11) becomes

$$y^2 + y = \gamma. \tag{8.12}$$

Thus by Theorem 8.4, (8.11) has solutions if and only if

$$\operatorname{Tr}\left(\frac{AC}{B^2}\right) = 0, \tag{8.13}$$

in which case the solutions are

$$x = \frac{B}{A}y, \quad \frac{B}{A}(y+1),$$

where y is a solution to (8.12). (Incidentally, the quantity AC/B^2 is called the *discriminant* of the equation (8.11). Whether or not the quadratic equation (8.11) has a root depends on a test applied to the discriminant: is the trace of the discriminant zero? In fields of characteristic $\neq 2$, the discriminant of (8.11) is $B^2 - 4AC$, and the appropriate test is this: does the discriminant have a square root?)

- Case 2. $B = 0$. Then with $\gamma = C/A$, (8.11) becomes

$$x^2 = \gamma. \tag{8.14}$$

In any characteristic other than 2, such an equation (with $\gamma \neq 0$) has either no roots or two roots; but in characteristic 2 there is always *exactly one root.* For example, with $\gamma = 1$, equation (8.14) becomes $x^2 = 1$, which has $x = 1$ as its only solution. More generally,

$$x = \gamma^{2^{n-1}}. \tag{8.15}$$

is the only solution to (8.14), as the following argument shows. Clearly (8.15) is a solution to (8.14) since $x^2 = \gamma^{2^n} = \gamma$. If however $y^2 = \gamma$ also then clearly $(x/y)^2 = 1$. But since $z^2 - 1 = (z-1)^2$ in $GF(2^n)$, the only possibility is that $x = y$.

In summary, the solution to the general quadratic equation (8.11) is as follows.

$$\text{If } B = 0, \quad \text{one solution}$$

$$\text{If } B \neq 0\text{: } \operatorname{Tr}\left(\frac{AC}{B^2}\right) = 1, \quad \text{no solutions}$$

$$\operatorname{Tr}\left(\frac{AC}{B^2}\right) = 0, \quad \text{two solutions.}$$

Finally, we note that even if $\mathrm{Tr}(AC/B^2) = 1$, all is not lost. For if we define $F = GF(2)$, $E = GF(2^n)$, $K = GF(2^{2n})$, then if $\gamma = AC/B^2$,

$$\begin{aligned}\mathrm{Tr}_F^K(\gamma) &= \mathrm{Tr}_F^E\left(\mathrm{Tr}_E^K(\gamma)\right)\\ &= \mathrm{Tr}_F^E(\gamma + \gamma^{2^n})\\ &= \mathrm{Tr}_F^E(0)\\ &= 0.\end{aligned}$$

Thus if $\mathrm{Tr}(\gamma) = 1$, the equation (8.11) will have two solutions in the quadratic extension field $GF(2^{2n})$.

Having spent the first part of this section discussing the theoretical properties of trace and norm, we conclude by giving an entirely practical application—the design of efficient circuits to perform multiplication in a the field $GF(2^m)$. These results are due to Elwyn Berlekamp.

Let $\{\alpha_0, \alpha_1, \ldots, \alpha_{m-1}\}$ be a basis for $GF(2^m)$ over $GF(2)$, i.e., a set of m elements of $GF(2^m)$, which are linearly independent over $GF(2)$. The corresponding *dual* basis is defined to be the unique set of elements

$$\{\beta_0, \beta_1, \ldots, \beta_{m-1}\} \subseteq GF(2^m)$$

such that

$$\text{(8.16)} \qquad \begin{aligned}\mathrm{Tr}(\alpha_i\beta_j) &= 1 \qquad \text{if } i = j\\ &= 0 \qquad \text{if } i \neq j.\end{aligned}$$

We omit the proof that the dual basis exists and is unique, but here is a mechanical procedure for finding it. Define the $m \times m$ matrix $A = (a_{ij})_{i,j=0}^{m-1}$ over $GF(2)$ as follows:

$$\text{(8.17)} \qquad a_{ij} = \mathrm{Tr}(\alpha_i\alpha_j),$$

and let $B = A^{-1}$. Then if the (k, j)th entry of B is denoted by b_{kj}, the dual basis is given by

$$\beta_j = \sum_{k=0}^{m-1} b_{kj}\alpha_k, \qquad \text{for } j = 0, \ldots, m-1. \tag{8.18}$$

The matrices A and B can be used to change coordinates from the α–basis (hereafter called the *primal* basis) to the β–basis (the *dual* basis). Thus suppose $x \in GF(2^m)$, and let

$$x = \sum_{i=0}^{m-1} x_i\alpha_i, \qquad x_i \in \{0, 1\} \tag{8.19}$$

be the primal basis expansion of x, and let

$$x = \sum_{j=0}^{m-1} x'_j\beta_j, \qquad x'_j \in \{0, 1\} \tag{8.20}$$

be the dual basis expansion. Then if we define the the binary column vectors $\mathbf{x}$ and $\mathbf{x}'$ by

$$\mathbf{x} = (x_0, x_1, \ldots, x_{m-1})^T \tag{8.21}$$

$$\mathbf{x}' = (x'_0, x'_1, \ldots, x'_{m-1})^T, \tag{8.22}$$

it is easy to check that

$$\mathbf{x}' = A\mathbf{x}, \qquad x'_j = \mathrm{Tr}(x\alpha_j) \tag{8.23}$$

$$\mathbf{x} = B\mathbf{x}', \qquad x_i = \mathrm{Tr}(x\beta_i). \tag{8.24}$$

Example 8.5. Consider the basis $\{1, \alpha, \alpha^2\}$ in $GF(8)$, where α is a primitive root which satisfies $\alpha^3 = \alpha + 1$. Then $\mathrm{Tr}(1) = 1$, $\mathrm{Tr}(\alpha) = \mathrm{Tr}(\alpha^2) = 0$. Thus if $x = (x_2, x_1, x_0)$ in primal coordinates, $\mathrm{Tr}(x) = x_0$. Here is a table:

i	α^i	$\mathrm{Tr}(\alpha^i)$
0	0 0 1	1
1	0 1 0	0
2	1 0 0	0
3	0 1 1	1
4	1 1 0	0
5	1 1 1	1
6	1 0 1	1

Hence the matrix A is given by

$$A = \begin{bmatrix} 1 & 0 & 0 \\ 0 & 0 & 1 \\ 0 & 1 & 0 \end{bmatrix} = \mathrm{Tr} \begin{bmatrix} \alpha^0 & \alpha^1 & \alpha^2 \\ \alpha^1 & \alpha^2 & \alpha^3 \\ \alpha^2 & \alpha^3 & \alpha^4 \end{bmatrix}.$$

In this case, we're lucky; A is a permutation matrix and $A^{-1} = A^T$:

$$B = \begin{bmatrix} 1 & 0 & 0 \\ 0 & 0 & 1 \\ 0 & 1 & 0 \end{bmatrix}.$$

Therefore the dual basis in this case is $\beta_0 = 1$, $\beta_1 = \alpha^2$, and $\beta_2 = \alpha$. ∎

In the rest of this chapter, we will consider only primal bases of the form $\{1, \alpha, \alpha^2, \ldots, \alpha^{m-1}\}$, where α is a primitive* element in $GF(2^m)$. We will as before denote the corresponding dual basis by $\{\beta_0, \beta_1, \ldots, \beta_{m-1}\}$. The key fact is that *multiplying x by α is easy in the dual coordinate system.*

To see why this is so, let x be expressed in dual coordinates as

$$x = x'_{m-1}\beta_{m-1} + x'_{m-2}\beta_{m-2} + \cdots + x'_0\beta_0. \tag{8.25}$$

---

* A *primitive element* (as distinct from a *primitive root*) is an element which does not lie in a subfield, i.e., such that $\{1, \alpha, \ldots, \alpha^{m-1}\}$ is a basis for $GF(2^m)/GF(2)$.

Then according to (8.23), we have

$$x'_j = \mathrm{Tr}(x \cdot \alpha^j), \qquad j = 0, 1, \ldots, m-1. \tag{8.26}$$

What are the components of αx in the dual coordinate system? By (8.23) again,

$$(\alpha x)'_j = \mathrm{Tr}(\alpha x \cdot \alpha^j) = \mathrm{Tr}(x \cdot \alpha^{j+1}). \tag{8.27}$$

Comparing (8.26) and (8.27), we see the following:

$$\begin{cases} (\alpha x)'_j = x'_{j+1} & j = 0, 1, \ldots, m-2 \\ (\alpha x)'_{m-1} = \mathrm{Tr}(x \cdot \alpha^m) & \end{cases} \tag{8.28}$$

The relationships in (8.28) suggest a simple shift-register that can multiply by α, which is shown in Figure 8.1. The "Parity Tree" in Figure 8.1 will, in any specific case, be a circuit that calculates a mod 2 sum of a subset of the contents of the m flip-flops corresponding to $\mathrm{Tr}(x\alpha^m)$. When the shift register is clocked, the *new* contents of the shift register are equal to the *old* contents multiplied by α. Everything is in dual coordinates.

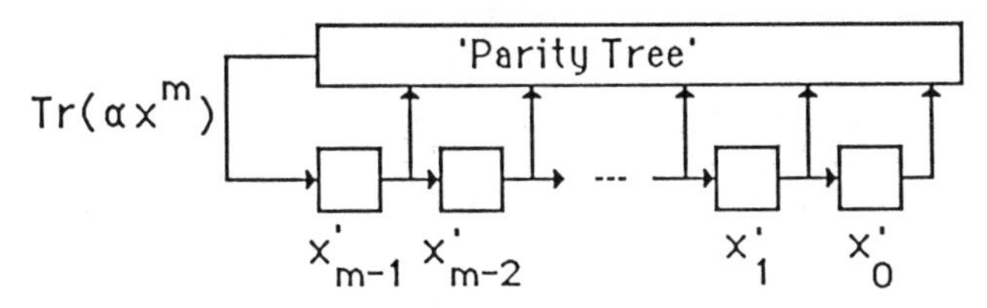

Figure 8.1. A generic "multiply by α" circuit.

Example 8.6. Continuing Example 8.5, we have, since $m = 3$, $\mathrm{Tr}(x \cdot \alpha^m) = \mathrm{Tr}(x\alpha^3)$. Since x is in dual coordinates, $x = x'_2\beta_2 + x'_1\beta_1 + x'_0\beta_0$, and $\alpha^3 = \alpha + 1$. Using the duality relationships (8.16), we get

$$\mathrm{Tr}(x\alpha^3) = x'_1 + x'_0,$$

and so the generic circuit in Figure 8.1 becomes, in this special case, the circuit in Figure 8.2. ∎

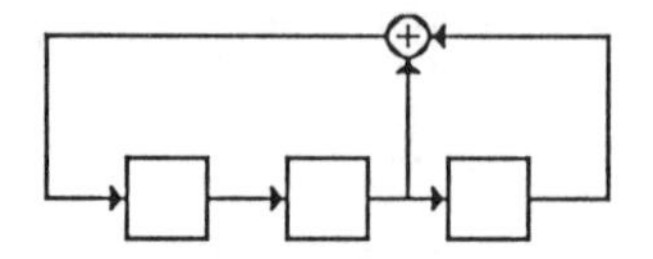

Figure 8.2. Multiplying by α in $GF(8)$.

So now we know how to multiply a *variable* x by the specific *constant* α. How can we multiply by other field elements? To see how, we look carefully at the contents of the ith flip-flop in Figure 8.1. We denote this by $\textbf{contents}_i(t)$. At $t = 0$, the ith flip-flop contains the ith bit of x, which, by (8.23) is $\mathrm{Tr}(\alpha^i x)$:

$$\textbf{contents}_i(0) = \mathrm{Tr}(\alpha^i x).$$

At time $t = 1$, the shift register contains αx, and so

$$\begin{aligned}\textbf{contents}_i(1) &= \mathrm{Tr}(\alpha^i \cdot \alpha x)\\ &= \mathrm{Tr}(\alpha \cdot \alpha^i x).\end{aligned}$$

In general, we have, for $0 \le t \le m-1$:

$$\begin{aligned}\textbf{contents}_i(t) &= \mathrm{Tr}(\alpha^i \cdot \alpha^t x)\\ &= \mathrm{Tr}(\alpha^t \cdot \alpha^i x).\end{aligned}$$

In other words,

$$\textbf{contents}_i(t) = t\text{th component of } \alpha^i x. \tag{8.29}$$

Thus if we tap the ith flip-flop, we will see the bits in $\alpha^i \cdot x$ appear *serially*; that is, after the tth clock cycle the ith flip-flop will contain the tth component of $\alpha^i \cdot x$ (dual coordinates). Since any element in $GF(2^m)$ can be written as a linear combination of $1, \alpha, \ldots, \alpha^{m-1}$, (8.29) tells us how to multiply x by an arbitrary constant μ (see Figure 8.3). The "Parity Tree 1" in Figure 8.3 is the same as in Figure 8.1; it is fixed by the choice of α. However, "Parity Tree 2" depends on the constant μ.

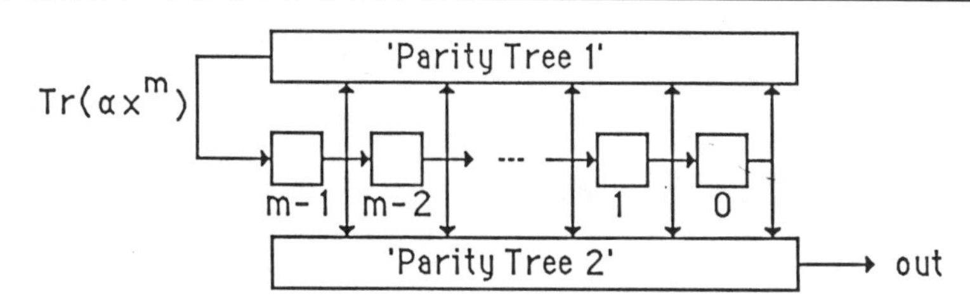

Figure 8.3. A generic bit serial "multiply by μ" circuit.

Example 8.7. Again consider $GF(8)$, with $\alpha^3 = \alpha + 1$. Suppose the constant μ is $\mu = \alpha^4$. From Example 8.6 we know $\alpha^4 = \alpha^2 + \alpha$, and so the general Figure 8.3 specializes to Figure 8.4. Initially the shift register in Figure 8.4 contains the dual coordinates of x; after t clock cycles, the output will be the tth component (again, in dual coordinates) of the product $\alpha^4 \cdot x$. ∎

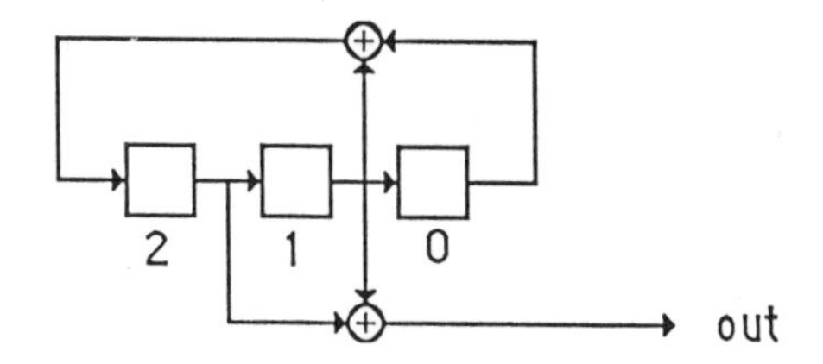

Figure 8.4. A circuit to perform bit-serial multiplication by α^4 in $GF(8)$.

So much for multiplying the *variable* x by the *constant* μ. What if we want to multiply the variable x by the *variable* y? Once again Eq. (8.29) is the key, because if y is expressed in *primal* coordinates as

$$y = y_{m-1}\alpha^{m-1} + y_{m-2}\alpha^{m-2} + \cdots + y_0 \cdot 1 \qquad (8.30)$$

then the following is true:

$$\begin{aligned}
&\sum_{i=0}^{m-1} y_i \cdot \mathbf{contents}_i(t) \\
&= \sum_{i=0}^{m-1} y_i \cdot (\alpha^i x)'_t \quad \text{(by Eq. (8.29))} \\
&= (x \cdot \sum y_i \alpha^i)'_t \\
&= (x \cdot y)'_t.
\end{aligned}$$

What this means is that if the primal coordinates of y are held in an m stage shift register, then after t clock cycles the dot product of the x and y registers will equal the tth bit of the product in *dual* coordinates. Thus Figure 8.5 is a generic variable-variable multiply circuit.

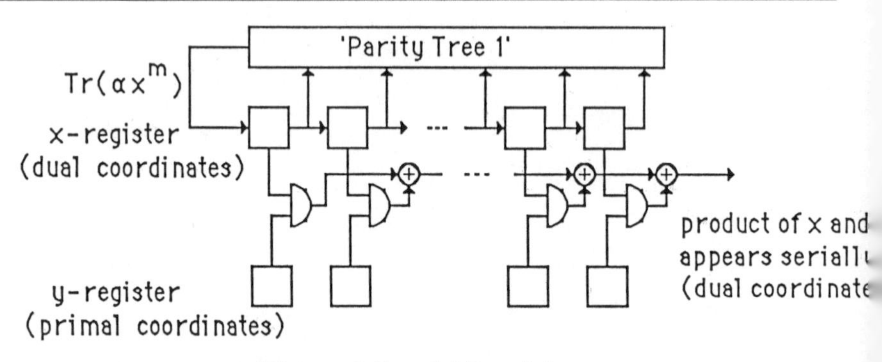

Figure 8.5. A bit-serial variable/variable multiplier in $GF(2^m)$.

In Figure 8.5, the x-register is initially loaded with the m bits $x'_{m-1}, \ldots, x'_0$ of the dual representation of x, and the y-register is loaded with the m bits $y_{m-1}, \ldots, y_0$ of the primal representation of y. After the tth clock cycle, the tth bit z'_t of the product xy expressed in *dual* coordinates appears at the output.

It is unfortunate that two different bases are involved in the circuit of Figure 8.5, and to actually use it as part of a larger device it would in general be necessary to have circuitry to change bases. However, sometimes this basis change is very easily accomplished. In Example 8.6, for example, we saw that in $GF(8)$ the basis change is just a permutation of the coordinates. We will exploit this fact in our final Example.

Example 8.8. Again $GF(8)$, $\alpha^3 = \alpha+1$, etc. In Figure 8.6 we have extended the circuit of Figure 8.2, according to the prescription of Figure 8.5, to produce a variable-variable bit serial multiplication circuit for $GF(8)$, in which both primal and dual coordinates are needed.

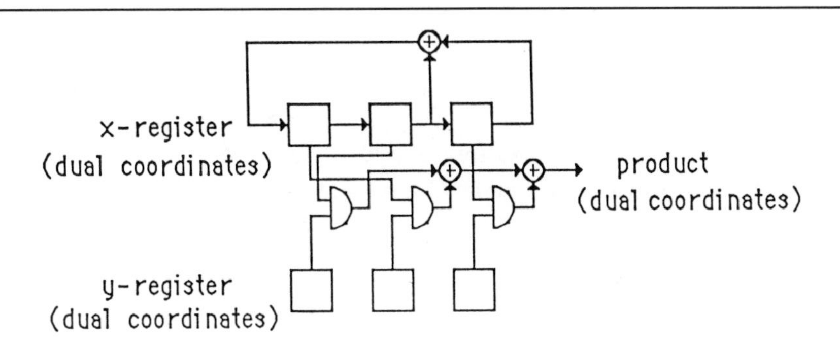

Figure 8.6. A variable/variable bit serial multiplier in $GF(8)$.

However, in Example 8.6 we saw that that to change y from dual coordinates (y'_2, y'_1, y'_0) to primal coordinates (y_2, y_1, y_0), we just permute:

$$y'_2 = y_1$$

$$y'_1 = y_2$$

$$y'_0 = y_0.$$

Thus the circuit of Figure 8.6 can be slightly rewired to give the simple bit-serial multiplier in Figure 8.7, where all coordinates are now dual coordinates. ∎

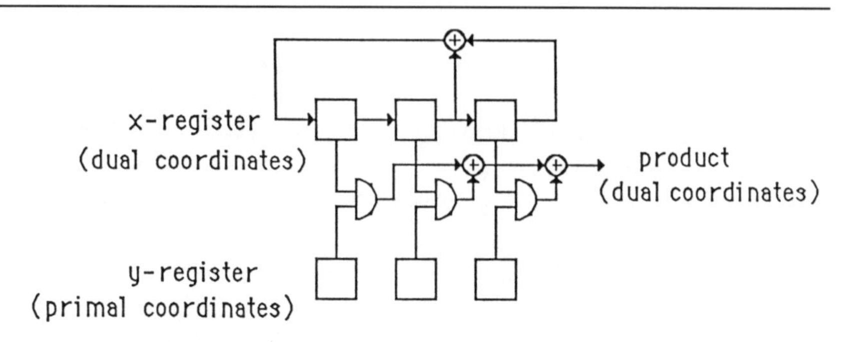

Figure 8.7. A variable/variable bit-serial multiply circuit for $GF(8)$. (All coordinates are dual coordinates.)

Of course, coordinate changing won't always be so easy. But Doug Whiting has shown that for $m \leq 12$, only $m = 8$ does not have this kind of "self-dual" basis. And even for $m = 8$ there is a basis-changing circuit that needs only three mod 2 adders. (Actually, to get these self-dual bases, Whiting enlarged the definition (8.16) of duality to

$$T(\alpha^i \beta_j) = \delta_{ij}, \tag{8.31}$$

where $T(x) = \text{Tr}(\lambda x)$ for some constant λ.)

Problems for Chapter 8.

1. Give the proofs for Norm in Theorem 8.1.

2. Construct the field $GF(16)$, using the basis $\{1, \alpha, \alpha^2, \alpha^3\}$, where α satisfies $\alpha^4 + \alpha^3 + 1 = 0$. What is the Trace of the element $\beta = b_0 + b_1\alpha + b_2\alpha^2 + b_3\alpha^3$?

3. Find a formula for the trace in $GF(32)$. (Use $x^5 + x^2 + 1$ as the defining polynomial.)

4. Let F and K be as in Theorem 8.1. If the elements in K are viewed as n dimensional vectors with components in F, *any* function of the form $T(\beta) = \beta \cdot c$ where c is a fixed vector will satisfy the properties (a), (b), (c), and (e) of Theorem 8.1. Show that the only such functions which also satisfy (d), i.e., $T(\beta) = T(\beta^q)$ are the functions of the form $T(\beta) = \lambda \operatorname{Tr}(\beta)$, $\lambda \in GF(q) - \{0\}$. (This result is due to Anselm Blumer.)

5. Give the complete proofs, for trace and norm, of Theorem 8.2.

6. Let $E = GF(q)$, $K = GF(q^m)$, and suppose that $\alpha \in E$.
a. Show that $\operatorname{Tr}_E^K(\alpha) = m \cdot \alpha$.
b. Show that $\operatorname{N}_E^K(\alpha) = \alpha^m$.

7. Prove Theorem 8.3 for norm.

8. State and prove a theorem like Theorem 8.3 which gives a formula for $\operatorname{Tr}_F^K(\alpha^{-1})$ and $\operatorname{N}_F^K(\alpha^{-1})$ in terms of the degree and the coefficients of the minimal polynomial of α.

9. Consider the field $GF(64)$, which contains a primitive root α satisfying $\alpha^6 + \alpha + 1 = 0$. Find an element θ with trace 1, and use it (via Lemma 8.5) to give a general solution to the quadratic equation (8.6), in the form given in (8.10).

10. Prove (8.23) and (8.24).

11. Consider the fields $F_m = GF(8^m)$, $m = 1, 2, \ldots$.
a. Show that there are exactly seven solutions to the equation $x^7 = 1$ in each such field.
b. Let a_m denote the number of solutions to $x^7 = 1$ with trace 0. Compute a_m, for all $m = 1, 2, \ldots$. (Here trace means the trace from F_m to $GF(2)$.)

12. Consider the field $GF(32)$, which contains a primitive root α satisfying $\alpha^5 = \alpha^2 + 1$. Find necessary and sufficient conditions on the element γ such that the equation

$$y^2 + y = \gamma$$

has two solutions in $GF(32)$. The conditions should be stated in terms of the coefficients γ_i in the expansion of γ with respect to the basis $\{1, \alpha, \alpha^2, \alpha^3, \alpha^4\}$ of $GF(32)$ (cf. Example 8.3).

13. In the field $GF(27)$, let α satisfy $\alpha^3 = \alpha + 2$. (α is a primitive root.)

a. Represent an arbitrary element γ as $\gamma = \gamma_2\alpha^2 + \gamma_1\alpha + \gamma_0$, with γ_0, γ_1, and γ_2 elements of the ground field $GF(3)$. Find a vector $\mathbf{t} = (t_2, t_1, t_0)$ such that $\mathrm{Tr}(\gamma) = \gamma_2 \cdot t_2 + \gamma_1 \cdot t_1 + \gamma_0 \cdot t_0$.

b. Find all solutions to the following three equations:

$$y^3 - y = \alpha^3$$
$$y^3 - y = \alpha^8$$
$$y^3 - y = 2\alpha + 2.$$

14. In this problem we will sketch Hilbert's constructive proof of the "norm" half of Theorem 8.4. Given elements α, θ in $GF(q^n)$, define

$$\beta = \theta + \alpha\theta^q + \alpha^{1+q}\theta^{q^2} + \cdots + \alpha^{1+q+\cdots+q^{n-2}}\theta^{q^{n-1}}.$$

a. Prove that $\alpha\beta^q + \theta = \beta + \mathrm{N}(\alpha) \cdot \theta$.

b. Prove that for any given α, there exists θ such that $\beta \neq 0$.

c. Finally show that if $\mathrm{N}(\alpha) = 1$, there exists a value of β such that $\alpha = \beta^{1-q}$.

15. For all 13 elements α of norm 1 in $GF(27)$, find the value of β such that $\alpha = \beta^{-2}$. (Use the primitive polynomial $x^3 + 2x + 1$ to generate the field.)

16. If $\{\alpha_0, \alpha_1, \ldots, \alpha_{m-1}\}$ is a basis for $GF(2^m)$ over $GF(2)$, show that a dual basis defined by (8.16) exists and is unique. [Hint: Use the matrices A and B defined in the text.]

17. Consider the basis $\{\alpha, \alpha^2, \alpha^3\}$ of $GF(8)$, where α is a primitive root which satisfies $\alpha^3 = \alpha + 1$. Find the matrices A and B as in Example 8.5, and find the dual basis.

18. Find a basis dual to $\{1, \alpha, \ldots, \alpha^{m-1}\}$ in these two fields:

a. $GF(32) : \alpha^5 = \alpha^2 + 1$.

b. $GF(64) : \alpha^6 = \alpha + 1$.

19. Design circuits to do bit-serial multiplication as follows:

a. Multiply an arbitrary field element in $GF(32)$ by α^{10}. [$\alpha^5 = \alpha^2 + 1$.]

b. Multiply two arbitrary elements in $GF(64)$. [$\alpha^6 = \alpha + 1$.]

Chapter 9

Linear Recurrences over Finite Fields

In this chapter we will study the general theory of linear recurrences over finite fields. However, we begin with a familiar example of a linear recurrence over the field of real numbers.

Let $(s_0, s_1, s_2, \ldots)$ be the sequence of real numbers defined as follows.

$$s_0 = 0, \quad s_1 = 1, \tag{9.1}$$

and

$$s_t = s_{t-1} + s_{t-2}, \qquad \text{for } t \geq 2. \tag{9.2}$$

Equation (9.2) is a called a *linear recurrence relation*, and the conditions in Eq. (9.1) are called *initial conditions.* Using (9.1) to get started, and (9.2) to continue, it is easy to compute as many terms of the sequence as we wish:

$$(s_0, s_1, s_2, \ldots) = (0, 1, 1, 2, 3, 5, 8, 13, 21, 34, 55, 89, \ldots). \tag{9.3}$$

These are the the famous *Fibonacci numbers.* They form the *particular solution* of the recurrence (9.2) satisfying the specific initial conditions given in (9.1). To find the *general* solution to (9.2), we proceed as follows.

We assume, as a guess, that there is a solution to (9.2) of the form

$$s_t = \alpha^t, \qquad t = 0, 1, 2, \ldots, \tag{9.4}$$

for some nonzero number α. Then (9.2) implies that

$$\alpha^t = \alpha^{t-1} + \alpha^{t-2}, \qquad \text{for } t \geq 2.$$

Dividing this by α^{t-2}, we find that α must satisfy the quadratic equation

$$\alpha^2 - \alpha - 1 = 0, \tag{9.5}$$

which is called the *characteristic equation* of the recurrence (9.2). There are of course two solutions to (9.5):

$$\alpha_1 = \frac{1+\sqrt{5}}{2}, \quad \alpha_2 = \frac{1-\sqrt{5}}{2}.$$

It follows that both of the sequences (α_1^t) and (α_2^t) satisfy (9.2), and indeed that any *linear combination* of these two sequences does, too. Thus if A and B are arbitrary constants, the sequence

$$s_t = A\left(\frac{1+\sqrt{5}}{2}\right)^t + B\left(\frac{1-\sqrt{5}}{2}\right)^t \tag{9.6}$$

will satisfy (9.2). Conversely, *every* solution to (9.2) is of the form (9.6). Rather than give a general proof of this, we illustrate its truth by determining the constants A and B for the Fibonacci numbers.

We are given in (9.1) that $s_0 = 0$, $s_1 = 1$. Thus if (9.6) is to yield the Fibonacci numbers, we must have

$$s_0 = A + B = 0$$

$$s_1 = A\left(\frac{1+\sqrt{5}}{2}\right) + B\left(\frac{1-\sqrt{5}}{2}\right) = 1.$$

One easily solves this system of equations and finds that $A = 1/\sqrt{5}$, $B = -1/\sqrt{5}$. Hence the tth Fibonacci number is given by the following remarkable explicit formula (we change notation and denote the tth Fibonacci number by F_t, as is usual in the literature)

$$F_t = \frac{1}{\sqrt{5}}\left(\frac{1+\sqrt{5}}{2}\right)^t - \frac{1}{\sqrt{5}}\left(\frac{1-\sqrt{5}}{2}\right)^t. \tag{9.7}$$

What we have done so far is well-known, but has nothing directly to do with finite fields. Indeed the whole subject of linear recurrences over the real or complex field is very well understood, and, we hope, somewhat familiar to our readers. In this chapter we will develop the theory of linear recurrences over finite fields, following the methods which apply to the real numbers as closely as possible. However, we will find that the structure of finite fields leads to some very nice properties which do not carry over to the reals or complexes.

Let us again consider the sequence (s_t) defined by (9.1) and (9.2), only now we assume that the s_t's lie not in the field of rational numbers, but in the finite field $GF(3) = \{0, 1, 2\}$. With this assumption, the sequence s_t turns out to be

$$0, 1, 1, 2, 0, 2, 2, 1, 0, 1, 1, 2, 0, \ldots . \tag{9.8}$$

This is the particular $GF(3)$ solution to the recurrence (9.2) corresponding to the initial conditions (9.1). This sequence can be viewed as the Fibonacci sequence (mod 3). Notice that it is *periodic* and has period 8.

Let us now attempt to find the *general* solution to (9.2) over the field $GF(3)$, using the same ideas we used when the underlying field was the field of rational numbers. Thus we again assume a solution of the form (9.4); this forces α to satisfy the characteristic equation (9.5), which over $GF(3)$ can be written as

$$f(x) = x^2 + 2x + 2 = 0. \tag{9.9}$$

Now $f(x)$ is *irreducible* over $GF(3)$, and so its two roots lie in the field $GF(9)$, which can be defined by using $f(x)$. Indeed if α is a root of (9.9), i.e., if

$\alpha^2 = \alpha + 1$, the field $GF(9)$ can be represented as follows:

i	α^i
0	1
1	α
2	$\alpha + 1$
3	$2\alpha + 1$
4	2
5	2α
6	$2\alpha + 2$
7	$\alpha + 2$

In this representation of $GF(9)$, the roots of the characteristic equation (9.9) are α and α^3, and so analogously to (9.6) we have a family of solutions to (9.2), viz.,

$$s_t = A\alpha^t + B\alpha^{3t}. \tag{9.10}$$

Let us see how to choose A and B so that the general formula (9.10) will produce the particular sequence (9.8). Again, we use the initial conditions (9.1):

$$s_0 = A + B = 0$$
$$s_1 = A\alpha + B\alpha^3 = 1.$$

From these two equations, we obtain $A = (\alpha + 1)$, $B = -(\alpha + 1)$:

$$s_t = (\alpha + 1)\alpha^t - (\alpha + 1)\alpha^{3t}, \tag{9.11}$$

which is an explicit formula, analogous to (9.7), for the Fibonacci numbers (mod 3). We can proceed further by noticing that $(\alpha + 1) = \alpha^2$, $-(\alpha + 1) = 2\alpha + 2 = \alpha^6$. Hence $-(\alpha + 1) = (\alpha + 1)^3$ and (9.11) becomes

$$\begin{aligned} s_t &= (\alpha + 1)\alpha^t + ((\alpha + 1)\alpha^t)^3 \\ &= \mathrm{Tr}((\alpha + 1)\alpha^t), \end{aligned} \tag{9.12}$$

where "Tr" denotes the trace from $GF(9)$ to $GF(3)$.

Much of the above discussion can be generalized. If (s_t) is a sequence of elements in a field F, and if $a_1, a_2, \ldots, a_m$ are fixed elements of F, a relationship of the form

$$s_t = a_1 s_{t-1} + a_2 s_{t-2} + \cdots + a_m s_{t-m} \tag{9.13}$$

is called an mth order linear recurrence relation over F. The *characteristic polynomial* of the recurrence (9.13) is defined to be

$$f(x) = x^m - a_1 x^{m-1} - a_2 x^{m-2} - \cdots - a_m. \tag{9.14}$$

The following theorem is the basic theorem of the subject. Its proof is left as an exercise, which should be an easy exercise, in light of the preceding discussion.

Theorem 9.1. *If $\alpha_1, \alpha_2, \ldots, \alpha_r$ are distinct roots of the characteristic polynomial $f(x)$, then for any choice of constants $\lambda_1, \lambda_2, \ldots, \lambda_r$ the sequence*

$$s_t = \lambda_1 \alpha_1^t + \lambda_2 \alpha_2^t + \cdots + \lambda_r \alpha_r^t$$

satisfies the linear recurrence (9.13). ∎

Most of the rest of this chapter will be devoted to exploring the consequences of Theorem 9.1 when the underlying field is the finite field $GF(q)$.

At first we will make the simplifying assumption that the associated characteristic polynomial is *irreducible*. Under this assumption, it follows that $f(x)$ has m distinct conjugate roots $\alpha, \alpha^q, \ldots, \alpha^{q^{m-1}}$ in $GF(q^m)$. Let $\mathrm{Tr}(x) = x + x^q + \cdots + x^{q^{m-1}}$ denote the trace from $GF(q^m)$ to $GF(q)$. The following theorem generalizes (9.12).

Theorem 9.2. *If the characteristic polynomial (9.14) is irreducible, then for any $\theta \in GF(q^m)$, the sequence (s_t) defined by*

$$s_t = \mathrm{Tr}(\theta \alpha^t), \tag{9.15}$$

satisfies the recurrence (9.13). Conversely, given a sequence (s_t) satisfying (9.13), there exists a unique $\theta \in GF(q^m)$ such that (9.15) holds.

Proof: Since α is a root of the equation $f(x) = 0$, where $f(x)$ is the characteristic polynomial (9.14), we have

$$\alpha^m = \sum_{i=1}^{m} a_i \alpha^{m-i}. \tag{9.16}$$

Thus if (s_t) is defined by (9.15), we have

$$\begin{aligned} s_t &= \mathrm{Tr}(\theta\alpha^{t-m}\alpha^m) \\ &= \mathrm{Tr}\left(\theta\alpha^{t-m} \cdot \sum_{i=1}^{m} a_i\alpha^{m-i}\right) \\ &= \sum_{i=1}^{m} \mathrm{Tr}\left(a_i\theta\alpha^{t-i}\right) \\ &= \sum_{i=1}^{m} a_i \,\mathrm{Tr}\left(\theta\alpha^{t-i}\right) \\ &= \sum_{i=1}^{m} a_i s_{t-i}. \end{aligned}$$

Thus (9.13) is satisfied.

To prove the converse, we need the following important lemma.

Lemma 9.3. *Let $\theta \in GF(q^m)$, and suppose α is such that $(1, \alpha, \ldots, \alpha^{m-1})$, are linearly independent over $GF(q)$. Then if*

$$\mathrm{Tr}(\theta\alpha^i) = 0 \qquad \text{for } i = 0, 1, 2, \ldots, m-1, \tag{9.17}$$

it follows that $\theta = 0$. Note that the hypothesis $(1, \alpha, \ldots, \alpha^{m-1})$, are linearly independent is equivalent to saying that the minimal polynomial of α has degree m (see Problem 9.3).

Proof: Let $\beta \in GF(q^m)$ be arbitrary. Since $(1, \alpha, \ldots, \alpha^{m-1})$ are linearly independent, they form a basis for $GF(q^m)$ with respect to the subfield $GF(q)$, and so there exist m elements of $GF(q)$, say $b_0, b_1, \ldots, b_{m-1}$, such that

$$\beta = \sum_{i=0}^{m-1} b_i \alpha^i.$$

Thus

$$\begin{aligned} \mathrm{Tr}(\theta\beta) &= \mathrm{Tr}\left(\sum_{i=0}^{m-1} \theta b_i \alpha^i\right) \\ &= \sum_{i=0}^{m-1} b_i \,\mathrm{Tr}(\theta\alpha^i) \\ &= 0 \qquad \text{(by (9.17))}. \end{aligned}$$

Hence $\mathrm{Tr}(\theta\beta) = 0$ for all β. If $\theta \neq 0$, this is impossible, since as β runs through all elements of $GF(q^m)$ so does $\theta\beta$, and by Theorem 8.1 we know that the trace isn't identically zero. ∎

We now continue with the proof of Theorem 9.2. It is clear that there are exactly q^m distinct solutions to the recurrence (9.13), corresponding to the q^m choices for the initial conditions $s_0, s_1, \ldots, s_{m-1}$. There are also q^m solutions to (9.13) of the form (9.15), corresponding to the q^m choices for θ. The proof of Theorem 9.2 will therefore be complete if we can show that the q^m solutions of the form (9.15) are all distinct. If two solutions of the form (9.15) *were* identical, we would have

$$\mathrm{Tr}(\theta_1\alpha^t) = \mathrm{Tr}(\theta_2\alpha^t), \qquad \text{for } t \geq 0.$$

This would imply that

$$\mathrm{Tr}((\theta_1 - \theta_2)\alpha^t) = 0, \qquad \text{for } t \geq 0.$$

But by Lemma 9.3, this means that $\theta_1 = \theta_2$. Thus the q^m solutions of the form (9.15) are distinct and so must account for all of the solutions to (9.13). ∎

We saw earlier that the Fibonacci numbers (mod 3) (see Eq. (9.8)), are periodic, of period 8. We shall now see that any linear recurring sequence over $GF(q)$ whose characteristic polynomial is irreducible is periodic, and learn how to calculate its period.

We say that the sequence (s_t) is periodic, of period n, if

$$s_{t+n} = s_t, \qquad \text{for all } t \geq 0. \tag{9.18}$$

If (s_t) is a solution to (9.13) and the characteristic polynomial is irreducible, then by Theorem 9.2, $s_t = \mathrm{Tr}(\theta\alpha^t)$ for some $\theta \in GF(q^m)$. If this sequence is to be periodic of period n, we must have

$$\mathrm{Tr}(\theta\alpha^{t+n}) = \mathrm{Tr}(\theta\alpha^t) \qquad t \geq 0, \text{ i.e.,}$$

$$\mathrm{Tr}(\theta\alpha^t(\alpha^n - 1)) = 0 \qquad t \geq 0.$$

Thus by Lemma 9.3 either $\theta = 0$ (i.e., $s_t = 0$ for all t), or else $\alpha^n = 1$. In the latter case we see that the sequence (s_t) is periodic of period n if and only if the order of α divides n. Hence the least n such that (9.18) holds is $\mathrm{ord}(\alpha)$. Now $\mathrm{ord}(\alpha) = n$ if and only if $f(x)$ divides the cyclotomic polynomial $\Phi_n(x)$. Thus we have proved the following theorem.

Theorem 9.4. *If $f(x)$ is irreducible, then every nonzero solution to (9.13) has (least) period n, where $n = \mathrm{ord}(\alpha)$. Equivalently, n is the least integer such that $x^n \equiv 1 \pmod{f(x)}$. Equivalently n is the unique integer such that $f(x) \mid \Phi_n(x)$. The integer n is called the* **period** *of $f(x)$.* ∎

Example 9.1. Let us again consider the Fibonacci number recursion (mod 3). Here $q = 3$ and $f(x) = x^2 - x - 1$. The polynomial $f(x)$ is irreducible, and so Theorems 9.2 and 9.4 apply. Indeed since $\Phi_8(x) = x^4 + 1$ factors as $(x^2 - x - 1)(x^2 + x - 1)$ over $GF(3)$, Theorem 9.4 tells us that every linear recurring sequence with characteristic polynomial $f(x)$ will have period 8. There are 9 such sequences, corresponding to the 9 possible choices for the initial values s_0 and s_1. These sequences can be listed in a 9×8 array as follows.

s_0	s_1	s_2	s_3	s_4	s_5	s_6	s_7
0	0	0	0	0	0	0	0
0	1	1	2	0	2	2	1
0	2	2	1	0	1	1	2
1	0	1	1	2	0	2	2
1	1	2	0	2	2	1	0
1	2	0	2	2	1	0	1
2	0	2	2	1	0	1	1
2	1	0	1	1	2	0	2
2	2	1	0	1	1	2	0

Let us now see how these 9 sequences can be produced by Theorem 9.2. Plainly the first sequence is produced by taking $\theta = 0$ in Theorem 9.2. To find the θ's associated with the other 8 sequences we need to work harder. If α is root of the equation $x^2 - x - 1 = 0$ in $GF(9)$, then α is a primitive root in $GF(9)$. It is easy to verify that $\mathrm{Tr}(1) = 2$ and $\mathrm{Tr}(\alpha) = 1$, and so the trace of an element expressed as $x_1\alpha + x_0$, where x_0 and x_1 are in $GF(3)$, is $x_1 + 2x_0$. Using these facts is it easy to produce the following table.

i	α^i	$\mathrm{Tr}(\alpha^i)$
0	1	2
1	α	1
2	$\alpha + 1$	0
3	$2\alpha + 1$	1
4	2	1
5	2α	2
6	$2\alpha + 2$	0
7	$\alpha + 2$	2

To identify the sequence beginning with $s_0 = 0$ and $s_1 = 1$, we need to find a value of θ with $\mathrm{Tr}(\theta) = 0$ and $\mathrm{Tr}(\theta\alpha) = 1$. Examining the trace table, we find that the only such θ is $\theta = \alpha^2$. Similarly, the sequence with $s_0 = 0$ and $s_1 = 2$ corresponds to $\theta = \alpha^6$. We leave the calculation of the θ corresponding to the remaining 6 sequences as a problem. ∎

If we examine the sequences listed in Example 9.1, we find a remarkable thing. Except for the first sequence, which is identically zero, each sequence is a cyclic shift of every other sequence! For example, the sequence beginning with $s_0 = 1$ and $s_1 = 0$ can be obtained from the sequence beginning with $s_0 = 2$ and $s_1 = 2$ by shifting left two digits (equivalently, by shifting right six digits). The same sort of thing happens in general; let us see why.

We define two solutions $(s_t), (s'_t)$ of (9.13) to be *cyclically equivalent*, if

$$s'_t = s_{t+i}, \qquad \text{for all } t \geq 0. \tag{9.19}$$

In terms of Theorem 9.2, this gives

$$\mathrm{Tr}(\theta'\alpha^t) = \mathrm{Tr}(\theta\alpha^{t+i}), \qquad t \geq 0$$

$$\mathrm{Tr}((\theta' - \theta\alpha^i)\alpha^t) = 0, \qquad t \geq 0.$$

By Lemma 9.3, it follows that (9.19) is equivalent to

$$\theta' = \theta\alpha^i. \tag{9.20}$$

Thus to find out how many cyclic equivalence classes of nonzero solutions to (9.13) there are, we must find out how many equivalence classes of nonzero elements there are in $GF(q^m)$, if equivalence is defined by (9.20). What (9.20) says, however, is that θ' and θ are equivalent if and only if they lie in the same coset of the subgroup $\{1, \alpha, \ldots, \alpha^{n-1}\}$ of the group of nonzero elements of $GF(q^m)$. Hence the number of equivalence classes equals the number of cosets, which is

$$N = \frac{q^m - 1}{n}. \tag{9.21}$$

We have proved the following.

Theorem 9.5. *If we define cyclic equivalence by (9.19), the number of cyclically inequivalent nonzero solutions to (9.13) is N, as defined by (9.21). In particular, if $N = 1$, i.e., if the characteristic polynomial is* **primitive**, *then all the nonzero solutions are cyclically equivalent.*

Example 9.2. Let $q = 3$, $f(x) = x^2 - x - 1$. Then as we have seen, $n = 8$, and so $N = (3^2 - 1)/8 = 1$. Thus $f(x)$ is primitive and all nonzero solutions to (9.13) are cyclically equivalent, a fact we have already noted. ∎

Example 9.3. Let $q = 2$, $f(x) = \Phi_5(x) = x^4 + x^3 + x^2 + x + 1$. By the results of Chapter 7, $\Phi_5(x)$ is irreducible over $GF(2)$, and so by Theorem 9.3 all solutions of the linear recurrence $s_t = s_{t-1} + s_{t-2} + s_{t-3} + s_{t-4}$ will have period $n = 5$. By Theorem 9.5, there are $N = (2^4 - 1)/5 = 3$ cyclic equivalence classes. In fact, the 16 solutions can be enumerated explicitly as follows:

	s_0	s_1	s_2	s_3	s_4		s_0	s_1	s_2	s_3	s_4
*	0	0	0	0	0	A	1	0	0	0	1
A	0	0	0	1	1	B	1	0	0	1	0
B	0	0	1	0	1	B	1	0	1	0	0
A	0	0	1	1	0	C	1	0	1	1	1
B	0	1	0	0	1	A	1	1	0	0	0
B	0	1	0	1	0	C	1	1	0	1	1
A	0	1	1	0	0	C	1	1	1	0	1
C	0	1	1	1	1	C	1	1	1	1	0

Apart from the all-zero solution, there are 15 solutions. They divide themselves into $N = 3$ equivalence classes of 5 elements each (these classes are labeled A, B, C), as predicted by Theorem 9.4. ∎

Example 9.4. Consider the recurrence

$$s_{t+2} = s_{t+1} - s_t \pmod 5. \tag{9.22}$$

The characteristic polynomial is $f(x) = x^2 - x + 1 = \Phi_6(x)$, which is irreducible over $GF(5)$ by the results of Chapter 7. All solutions to (9.22) therefore have period $n = 6$, and by Theorem 9.4 there are $N = (5^2 - 1)/6 = 4$ cyclic equivalence classes of nonzero solutions to (9.22). If we represent $GF(5)$ by $\{0, \pm 1, \pm 2\}$ these inequivalent solutions, which we will call *cycles*, are:

$$A : (0, 1, 1, 0, -1, -1)$$
$$B : (1, 2, 1, -1, -2, -1)$$
$$C : (0, 2, 2, 0, -2, -2)$$
$$D : (2, -1, 2, -2, 1, -2).$$

Notice that cycles A and C, though not cyclic shifts of each other, are nevertheless simply related, viz., C is A multiplied by the field element 2. B and D are similarly related. Thus in fact there are only two "really" distinct nonzero solutions to the recurrence (9.22), viz., A and B. Let us now study this phenomenon in general, for irreducible $f(x)$'s. ∎

We define two solutions $(s'_t), (s_t)$ to (9.13) to be ***projectively cyclically equivalent*** provided there exists an integer i and a nonzero scalar $\lambda \in GF(q)$ such that

$$s'_t = \lambda s_{t+i} \qquad \text{for all } t \geq 0. \tag{9.23}$$

We wish to count the number of inequivalent solutions to (9.13), using (9.23) as the definition of equivalence.

By Theorem 9.2, (9.23) can be written as

$$\mathrm{Tr}(\theta' \alpha^t) = \mathrm{Tr}(\lambda \theta \alpha^{t+i}), \qquad t \geq 0, \text{ or}$$

$$\mathrm{Tr}((\theta' - \lambda \theta \alpha^i)\alpha^t) = 0. \tag{9.24}$$

Hence by Lemma 9.3, it follows that (9.23) is equivalent to

$$\frac{\theta'}{\theta} = \lambda \alpha^i \tag{9.25}$$

The set of elements in $GF(q^m)$ of the form $\lambda\alpha^i$ where λ is a nonzero element of $GF(q)$ is a subgroup of the multiplicative group of nonzero elements of $GF(q^m)$. What (9.25) says is that θ' and θ are equivalent if and only if θ and θ' lie in the same coset of this subgroup. It follows that the number of inequivalent θ's is equal to the number of such cosets, viz.

$$N_1 = \frac{q^m - 1}{|G|}, \tag{9.26}$$

where G is the subgroup of elements of the form $\{\lambda\alpha^i\}$. It remains to calculate $|G|$. Now G is the direct product of the two groups L and A:

$$L = GF(q) - \{0\}, \quad A = \{1, \alpha, \ldots, \alpha^{n-1}\},$$

which are both subgroups of $GF(q^m) - \{0\}$. From elementary group theory we have

$$|G| = |A| \cdot |L| \;/\; |A \cap L|. \tag{9.27}$$

Plainly $|A| = n$ and $|L| = q-1$. To calculate $|A \cap L|$ we note that this number is just the number of distinct powers of α which are elements of $GF(q)$. But $\alpha^i \in GF(q)$ if and only if $\alpha^{i(q-1)} = 1$. Since $\text{ord}(\alpha) = n$, this is equivalent to

$$n \mid i(q-1), \quad \text{i.e.}$$

$$\frac{n}{\gcd(n, q-1)} \mid i.$$

Thus if we define

$$e = \gcd(n, q-1) \tag{9.28}$$

$$d = n/e, \tag{9.29}$$

we see that $\alpha^i \in GF(q)$ iff $i = 0, d, 2d, \ldots, (e-1)d$. Hence $|A \cap L| = e$, and so by (9.27) we have

$$|G| = n(q-1)/e,$$

and so by (9.26),

$$N_1 = \frac{q^m - 1}{n(q-1)} \cdot e. \tag{9.30}$$

It is an easy exercise to prove that an equivalent formulation is

$$N_1 = \gcd\left(N, \frac{q^m - 1}{q - 1}\right). \tag{9.31}$$

This then is the number of projectively, cyclically, inequivalent (in the sense of (9.23)) solutions to (9.13). Let us return to our previous example.

Example 9.4 (continued.) We recall that $q = 5$ and $n = 6$. Hence by (9.28) $e = \gcd(6, 4) = 2$, and so by (9.30) $N_1 = (5^2 - 1/6 \cdot 4) \cdot 2 = 2$. This confirms the observation we made in Example 9.4 that there are in fact only two "really" distinct solutions to the recurrence (9.22):

$$A = (0, 1, 1, 0, -1, -1)$$
$$B = (1, 2, 1, -1, -2, -1)$$

■

Next we draw your attention to the fact in Example 9.4, the solution A, which has period 6, "almost" has period 3, since

$$s_{t+3} = -s_t \qquad \text{for all } t.$$

The same thing holds for solution B. This phenomenon is no accident; the general rule is that

$$s_{t+d} = \lambda s_t \qquad \text{for all } t, \tag{9.32}$$

where d is as defined in (9.29), and λ is a nonzero element of $GF(q)$ satisfying $\lambda^e = 1$. This result follows directly from the basic Theorem 9.2 and the facts that

$$\alpha^d \in GF(q) \quad \text{and} \quad (\alpha^d)^e = 1. \tag{9.33}$$

In this context d is usually called the "reduced period" of (s_t) and λ is called a *multiplier.* Summarizing, we have proved the following.

Theorem 9.6. *If we define projective cyclic equivalence by (9.23), the number of projectively cyclically equivalent nonzero solutions to (9.13) is N_1, defined by (9.30) or (9.31). Furthermore, every nonzero solution to (9.13) satisfies (9.32), where the reduced period d is given by (9.29). The multiplier λ in (9.33) is an element of $GF(q)$ of order e, where e is given by (9.28).*

We now begin a brief study of a class of problems which can be called *distribution problems.* Let $(s_t)_{t=0}^{\infty}$ be a solution to the recurrence (9.13) with characteristic polynomial $f(x)$. Suppose that (s_t) has (least) period n. We define, for each $\alpha \in GF(q)$, the number

$$n_a(s) = |\{t : 0 \leq t \leq n-1, \quad s_t = a\}| . \tag{9.34}$$

Thus $n_a(s)$ denotes the number of times the element a occurs in one period of s. These numbers tell us how the elements of $GF(q)$ are *distributed* in the solutions to the recurrence relation (9.13). It is in general difficult to obtain information about these numbers; but in the following theorem n_0 is identified in an important special case.

Theorem 9.7. *If $f(x)$ is irreducible and if $N_1 = 1$ (see (9.30) or (9.31)), then*

$$n_0(s) = \frac{(q^{m-1} - 1)}{N},$$

for all solutions to (9.13) except the all zero solution.

Proof: Let $\theta_1, \theta_2, \ldots, \theta_N$ be a complete set of representatives for the cosets of $\{1, \alpha, \ldots, \alpha^{n-1}\}$ in the multiplicative group $GF(q^m)$. Every nonzero element

$\theta \in GF(q^m)$ can be written as $\theta = \theta_i \alpha^j$ for a unique pair (i, j), $1 \le i \le N$, $0 \le j \le n-1$. Now consider the following $N \times n$ array, which we call Array 1:

$$\text{(Array 1)} \qquad \begin{array}{ccccc} \theta_1 & \theta_1\alpha & \theta_1\alpha^2 & \cdots & \theta_1\alpha^{n-1} \\ \theta_2 & \theta_2\alpha & \theta_2\alpha^2 & \cdots & \theta_1\alpha^{n-1} \\ \vdots & \vdots & \vdots & & \vdots \\ \theta_N & \theta_N\alpha & \theta_N\alpha^2 & \cdots & \theta_N\alpha^{n-1} \end{array}$$

Now let $s_{ij} = Tr(\theta_i \alpha^j)$ and consider this array, which we call Array 2 :

$$\text{(Array 2)} \qquad \begin{array}{ccccc} s_{10} & s_{11} & s_{12} & \cdots & s_{1,n-1} \\ s_{20} & s_{21} & s_{22} & \cdots & s_{2,n-1} \\ \vdots & \vdots & \vdots & & \vdots \\ s_{N0} & s_{N1} & s_{N2} & \cdots & s_{N,n-1} \end{array}$$

By Theorems 9.2 and 9.5, every nonzero solution to (9.13) is (cyclically equivalent to) one of the rows of Array 2. But since Array 2 is the "trace" of Array 1, and since every nonzero element of $GF(q^m)$ appears exactly once in Array 1, it follows that 0 appears exactly $q^{m-1} - 1$ times in Array 2. Finally, since $N_1 = 1$, we know that every row of Array 2 can be obtained from the first row by shifting and multiplying by scalars. Thus 0 appears the same number of times in each row of Array 2. Since there are N rows in the array, and 0 appears $q^{m-1} - 1$ times altogether, each row contains exactly $(q^{m-1} - 1)/N$ zeros. ∎

Example 9.5. Let $q = 7$ and $f(x) = x^2 - x - 1$, i.e., the Fibonacci recursion over $GF(7)$. The polynomial $f(x)$ is irreducible, as is easily verified. Assuming the initial condition $s_0 = 0, s_1 = 1$, the solution to the recursion (9.13) is

$$A: 0, 1, 1, 2, 3, 5, 1, 6, 0, 6, 6, 5, 4, 2, 6, 1\,[0, 1, \ldots].$$

Thus the period is $n = 16$. A simple calculation gives $N = 3$, $e = 2$, $d = 8$, $N_1 = 1$. Thus Theorem 9.7 applies and it predicts that $n_0(2) = (7-1)/3 = 2$ for all $s \neq 0$. And we see that this is true, since 0 occurs twice among the 16 elements of the above period, and every other solution is a shift and/or a scalar multiple of the one given. Indeed, the other two cyclically inequivalent solutions are:

$$B : (0, 2, 2, 4, 6, 3, 2, 5, 0, 5, 5, 3, 1, 4, 5, 2)$$
$$C : (0, 3, 3, 6, 2, 1, 3, 4, 0, 4, 4, 1, 5, 6, 4, 3).$$

The following table gives the values of $n_a(s)$ for all nonzero s's. These numbers depend only on which of the 3 solutions (A, B, or C), the given solution is cyclically equivalent to.

	A	B	C
n_0	2	2	2
n_1	4	1	2
n_2	2	4	1
n_3	1	2	4
n_4	1	2	4
n_5	2	4	1
n_6	4	1	2

Note that only n_0 is independent of the particular equivalence class, although certain patterns among the other n_a's emerge. ($n_a = n_{-a}$, each n_a is a power of 2; etc.) ∎

All of our results so far needed to assume that the characteristic polynomial of the recurrence relation in question was irreducible. Let us now address the question of what happens when the characteristic polynomial is reducible. The following theorem settles the simplest case.

Theorem 9.8. *Suppose $f(x)$ factors completely into linear factors in $GF(q)$, i.e.,*

$$f(x) = \prod_{k=1}^{m} (x - \alpha_k), \qquad \alpha_k \in GF(q).$$

Furthermore assume the roots α_k are all distinct. Then every sequence of the form

$$s_t = \sum_{k=1}^{m} \theta_k \alpha_k^t, \qquad \theta_k \in GF(q) \tag{9.35}$$

satisfies the recurrence (9.13). Conversely, given a sequence (s_t) satisfying (9.13), there exists a unique m-tuple $(\theta_1, \theta_2, \ldots, \theta_m)$, of elements from $GF(q)$, such that (9.35) holds.

Proof: That any (s_t) defined by (9.35) satisfies (9.13) follows from Theorem 9.1.

To prove that every solution to (9.13) has the form (9.35), we observe that there are exactly q^m solutions to (9.13), corresponding to the q^m choices for the initial conditions $(s_0, s_1, \ldots, s_{m-1})$. There are also q^m solutions to (9.13) of the form (9.35). The following lemma will enable us to show that these solutions are all distinct.

Lemma 9.9. *Let $\alpha_1, \alpha_2, \ldots, \alpha_m$ be distinct elements of a field F. Then the only solution to the m simultaneous linear equations*

$$\sum_{k=1}^{m} x_k \alpha_k^t = 0, \qquad t = 0, 1, \ldots, m-1$$

is $x_1 = x_2 = \cdots = x_m = 0$.

Proof: In matrix form these equations are

$$\begin{pmatrix} 1 & 1 & \cdots & 1 \\ \alpha_1 & \alpha_2 & \cdots & \alpha_m \\ \alpha_1^2 & \alpha_2^2 & \cdots & \alpha_m^2 \\ \vdots & & & \vdots \\ \alpha_1^{m-1} & \alpha_2^{m-1} & \cdots & \alpha_m^{m-1} \end{pmatrix} \begin{pmatrix} x_1 \\ x_2 \\ \cdots \\ \cdots \\ \cdots \\ x_n \end{pmatrix} = \begin{pmatrix} 0 \\ 0 \\ \cdots \\ \cdots \\ \cdots \\ 0 \end{pmatrix}.$$

This matrix is a *Vandermonde* matrix and its determinant is known to be $\prod_{1 \le i \le j < m} (\alpha_j - \alpha_i)$; since the α_k's are by hypothesis distinct, the determinant is nonzero, and so the matrix is nonsingular. Thus the equations admit only the all-zero solution. ∎

We now resume the proof of Theorem 9.8. If two solutions (s_t) and (s'_t) of the form (9.35) happened to be identical, we would have

$$\sum_{k=1}^{m} \theta_k \alpha_k^t = \sum_{k=1}^{m} \theta'_k \alpha_k^t \qquad \text{for all } t \geq 0, \text{ i.e.,}$$

$$\sum_{k=1}^{m} (\theta_k - \theta'_k) \alpha_k^t = 0 \qquad \text{for all } t \geq 0.$$

Therefore by Lemma 9.9 we must have $\theta_k = \theta'_k$ for $k = 1, \ldots, m$. Thus the q^m apparently distinct solutions to (9.13) of the form (9.35) are in fact distinct, and so account for all of the q^m solutions to (9.13). ∎

Example 9.6. Let $q = 11, f(x) = x^2 - x - 1$, i.e., the Fibonacci recursion over $GF(11)$. Then $f(x) = (x - 4)(x - 8)$, and so by Theorem 9.8, every solution to the Fibonacci recursion (mod 11) is of the form $s_t = \theta_1 4^t + \theta_2 8^t$ (mod 11). For $s_0 = 0, s_1 = 1$, we find that $\theta_1 = 8$, $\theta_2 = 3$, and so the tth Fibonacci number F_t satisfies

$$F_t \equiv 8 \cdot 4^t + 3 \cdot 8^t \pmod{11}.$$ ∎

We continue our discussion of what happens when the characteristic polynomial factors into linear factors over $GF(q)$, and consider the question of periodicity. The sequence (s_t) has period n if and only if

$$s_{t+n} = s_t, \qquad \text{for } t = 0, 1, \ldots. \tag{9.36}$$

In terms of the results of Theorem 9.8, (9.36) says that

$$\sum_{k=1}^{m} \theta_k \alpha_k^{t+n} = \sum_{k=1}^{m} \theta_k \alpha_k^t$$

$$\sum_{k=1}^{m} (\theta_k \alpha_k^n - \theta_k) \alpha_k^t = 0 \qquad \text{for all } t \geq 0.$$

Thus by Lemma 9.9, (9.36) is equivalent to

$$\theta_k(\alpha_k^n - 1) = 0, \qquad \text{for } k = 1, 2, \ldots, m. \tag{9.37}$$

Equation (9.37) can be used to compute the period of (s_t), as follows. Define

$$n_k = \operatorname{ord}(\alpha_k), \qquad \text{for } k = 1, 2, \ldots, m. \tag{9.38}$$

Then if we define the subset K of $\{1, 2, \ldots, m\}$ by

$$K = \{k : \theta_k \neq 0\}, \tag{9.39}$$

it follows from (9.37) that $s_{t+n} = s_t$ for all t if and only if $\alpha_k^n = 1$ for all $k \in K$, i.e., n must be a multiple of $\operatorname{lcm}\{n_k : k \in K\}$. We state this result as a theorem.

Theorem 9.10. *The period of the sequence (s_t) given by (9.35) is*

$$n = \operatorname{lcm}\{n_k : k \in K\},$$

where $\{n_k\}$ and K are given by (9.38) and (9.39).

Example 9.7. We saw in Example 9.5 that the Fibonacci numbers F_t satisfy $F_t \equiv 8 \cdot 4^t + 3 \cdot 8^t \pmod{11}$. In the notation of Theorem 9.10, we have $\alpha_1 = 4$, $\alpha_2 = 8$, $\theta_1 = 8$, $\theta_2 = 3$, $K = \{1, 2\}$. Now $n_1 = \operatorname{ord}(\alpha_1) = \operatorname{ord}(4) = 5$, because $4^5 \equiv 1 \pmod{11}$; $n_2 = \operatorname{ord}(\alpha_2) = \operatorname{ord}(8) = 10$, since $8^{10} \equiv 1 \pmod{11}$. Thus $n = \operatorname{lcm}(5, 10) = 10$, and indeed the Fibonacci numbers $\pmod{11}$ are

$$0, 1, 1, 2, 3, 5, 8, 2, 10, 1, [0, 1, 1, 2, \ldots].$$

The general solution to the Fibonacci recursion (mod 11) is by Theorem 9.8

$$s_t = \theta_1 \cdot 4^t + \theta_2 8^t.$$

By Theorem 9.10, the period of this solution will be

$$\begin{array}{rll} 1 & \text{if} \quad \theta_1 = \theta_2 = 0 & \text{(one case)} \\ 5 & \text{if} \quad \theta_1 \neq 0, \theta_2 = 0 & \text{(10 cases)} \\ 10 & \text{if} \quad \theta_2 \neq 0 & \text{(110 cases).} \end{array}$$

This accounts for all $121 = 11^2$ possible values of the initial conditions s_0 and s_1. ∎

Let us now proceed to the general case of a characteristic polynomial with no repeated factors, but with no further restrictions. We assume that $f(x)$ factors as follows.

$$f(x) = \prod_{k=1}^{r} f_k(x), \qquad \deg(f_k) = m_k \tag{9.40}$$

where the f_k's are distinct and irreducible over $GF(q)$. By Theorem 6.1, we know that each $f_k(x)$ factors in $GF(q^{m_k})$ as follows:

$$f_k(x) = \prod_{j=0}^{m_k - 1} \left(x - \alpha_k^{q^j}\right), \qquad \alpha_k \in GF(q^{m_k}) \tag{9.41}$$

If we define the number M as $M = \operatorname{lcm}\{m_1, m_2, \ldots, m_r\}$, then each $GF(q^{m_k})$ will be a subfield of $GF(q^M)$ (Theorem 6.6) and the factorizations (9.41) will all hold simultaneously in $GF(q^M)$, and so in the big field $GF(q^M)$, $f(x)$ factors into linear factors. It follows then from Theorem 9.8 that given a sequence (s_t) from $GF(q)$ that satisfies the recursion with characteristic polynomial $f(x)$, (this will also be a sequence from $GF(q^M)$) we will have

$$s_t = \sum_{k=1}^{r} \sum_{j=0}^{m_k - 1} \theta_{k,j} \alpha_k^{q^j t}, \tag{9.42}$$

for certain constants $\theta_{k,j} \in GF(q^M)$. The problem with this is that not every such sequence will belong to $GF(q)$. To solve this problem, we use

Lemma 5.10, which implies that if $s_t \in GF(q)$, then $s_t^q = s_t$. Taking the expression for s_t in (9.42) to the qth power, we find:

$$s_t^q = \sum_{k=1}^{r} \sum_{j=0}^{m_k-1} \theta_{k,j}^q \alpha_k^{q^{j+1}t}. \tag{9.43}$$

According to Lemma 9.9, the only way that the two expressions (9.42) and (9.43) can be equal for all t is if the coefficients of the various roots of $f(x)$ are the same, i.e.,

$$\theta_{k,j+1} = \theta_{k,j}^q \qquad \text{for all } k, j.$$

Thus $\theta_{k,1} = \theta_{k,0}^q$, $\theta_{k,2} = \theta_{k,1}^q = \theta_{k,0}^{q^2}$, etc. Therefore if we define θ_k to be $\theta_{k,0}$, we conclude that the inner sum in (9.42) is in fact

$$\theta_k \alpha^t + \theta_k^q \alpha^{q^t} + \cdots + \theta_k^{q^{m_k-1}} \alpha^{q^{m_k-1}}.$$

But this is just the trace from $GF(q^{m_k})$ to $GF(q)$ of $\theta_k \alpha^t$. In summary:

Theorem 9.11. *The general solution to the recursion (9.13) when the characteristic polynomial factors as in (9.39) and (9.40) is*

$$s_t = \sum_{k=1}^{r} \mathrm{Tr}^{(k)} \left(\theta_k \alpha_k^t\right), \qquad \theta_k \in GF\left(q^{m_k}\right), \tag{9.44}$$

where $f_k(\alpha_k) = 0$, and $\mathrm{Tr}^{(k)}$ denotes the trace from $GF(q^{m_k})$ to $GF(q)$.

Corollary 9.12. *The period of the sequence (9.44) is*

$$n = \mathrm{lcm}(n_k : k \in K),$$

where n_k and K are defined as in (9.38) and (9.39).

Example 9.8. Consider the recursion

$$s_t = 3s_{t-1} + 2s_{t-2} + 2s_{t-3} \pmod 5,$$

whose characteristic polynomial factors as follows.

$$\begin{aligned} f(x) &= x^3 - 3x^2 - 2x - 2 \\ &= (x^2 - x + 1)(x - 2). \end{aligned}$$

The first factor is irreducible (in fact it is $\Phi_6(x)$), and so if α denotes a 6th root of unity in $GF(25)$, and if Tr denotes the trace in $GF(25)$, according to Theorem 9.11 the general solution to the given recursion is

$$s_t = \mathrm{Tr}(\theta_1 \alpha^t) + \theta_2 2^t,$$

Where θ_1 can assume any value in $GF(25)$ and θ_2 can assume any value in $GF(5)$. Here $n_1 = \mathrm{ord}(\alpha) = 6$, and $n_2 = \mathrm{ord}(2) = 4$, and so by Corollary 9.12, the period of such a sequence is

$$\begin{array}{rlll} 1 & \text{if } \theta_1 = 0, & \theta_2 = 0 & \text{(1 case)} \\ 4 & \text{if } \theta_1 = 0, & \theta_2 \neq 0 & \text{(4 cases)} \\ 6 & \text{if } \theta_1 \neq 0, & \theta_2 = 0 & \text{(24 cases)} \\ 12 & \text{if } \theta_1 \neq 0, & \theta_2 \neq 0 & \text{(96 cases).} \end{array}$$

This accounts for all $125 = 25^2$ possible values of the initial conditions s_0 and s_1. ∎

We conclude this chapter with some very brief remarks about what happens when the characteristic polynomial has repeated roots. We consider only the simplest case, in which the characteristic polynomial is a perfect power of a monomial,

$$f(x) = (x - \alpha)^m, \qquad \text{for some } \alpha. \tag{9.45}$$

Since α is a root of the characteristic equation, we know by Theorem 9.1 that $s_t = \lambda\alpha^t$, is a solution to the corresponding recursion for any $\lambda \in GF(q)$. This accounts for q solutions. But we know there must be q^m solutions in all. Where are the others? It turns out that, in fact, over any field F, finite or not, any sequence of the form

$$s_t = P(t)\alpha^t \tag{9.46}$$

is a solution for any function $P(t)$ mapping the integers into F, provided only that

$$\Delta^m P(t) = 0, \qquad \text{for all } t \geq 0. \tag{9.47}$$

In (9.47), the difference operator Δ is defined as follows:

$$\Delta P(t) = P(t+1) - P(t)$$
$$\Delta^k P(t) = \Delta(\Delta^{k-1} P(t)).$$

If P is a *real-valued* function, one can easily show that (9.47) holds if and only if P is a polynomial in t of degree $\leq m-1$. But over a finite field, this is not true if $m \geq p$ (the characteristic of the field). For example, if $q = 2$, the function $P(t)$ defined by

$$P(t) = \begin{cases} 0 & \text{if } t \equiv 0, 1 \pmod 4 \\ 1 & \text{if } t \equiv 2, 3 \pmod 4 \end{cases}$$

is not a polynomial of any degree, although $\Delta^3 P(t) = 0$.

The way out of this difficulty is the observation that the *binomial coefficients* can be used to build up the desired functions. In a general field, the binomial coefficients are defined recursively, as follows.

$$\binom{t}{0} = 1 \qquad \text{for all } t \geq 0, \text{ and}$$

$$\binom{t}{k} = \binom{t}{k-1} + \binom{t-1}{k-1} \qquad \text{for } t \geq 0 \text{ and } k \geq 1.$$

It is possible to show that for any field F, $\Delta^m P(t) = 0$ if and only if $P(t)$ can be written in the form

$$P(t) = \sum_{k=0}^{m-1} \lambda_k \binom{t}{k},$$

for scalars $\lambda_k \in F$. We have therefore proved the following theorem.

Theorem 9.13. *The general solution to the recurrence whose characteristic polynomial factors as in (9.45) is*

$$s_t = \left(\sum_{k=0}^{m-1} \lambda_k \binom{t}{k} \right) \alpha^t,$$

where $\lambda_1, \lambda_2, \ldots, \lambda_m$ are arbitrary elements of F. ∎

Example 9.9. Consider the recurrence

$$s_t = s_{t-1} + s_{t-2} + s_{t-3}$$

over $GF(2)$. The characteristic polynomial is $x^3 + x^2 + x + 1$, which factors as $(x+1)^3$. This is of the form (9.45) and so according to Theorem 9.13, the general solution to the given recurrence relation is

$$s_t = \lambda_0 + \lambda_1 t + \lambda_2 \binom{t}{2},$$

where each λ_i is either 0 or 1. ∎

Problems for Chapter 9.

1. Show that every solution to (9.2) is of the form (9.6).

2. Prove Theorem 9.1.

3. The hypothesis of Lemma 9.3 is that $(1, \alpha, \ldots, \alpha^{m-1})$ are linearly independent over $GF(q)$. Show that this hypothesis is equivalent to either one of the following:

a. The minimal polynomial of α has degree m.

b. α does not lie in a subfield of $GF(q^m)$.

4. Consider the sequence of integers $S_m = 2^m - 1$, for $m = 0, 1, 2, \ldots$, i.e., the sequence $0, 1, 3, 7, \ldots$. Let p be an odd prime, and consider the sequence (s_m) obtained by reducing the sequence (S_m) mod p, as a sequence of elements in $GF(p)$.

a. Show that the sequence (s_m) satisfies a linear recurrence over $GF(p)$, and find the characteristic polynomial.

b. For which values of p is the characteristic polynomial found in part (a) irreducible?

c. Find an explicit formula for s_m in the field $GF(11)$.

5. For each of the 9 sequences listed in Example 9.1, find the corresponding $\theta \in GF(9)$ such that $s_t = \mathrm{Tr}(\theta\alpha^t)$. (Note: Three of these θ's were already found in Example 9.1.)

6. Let α be a primitive root in $GF(16)$ satisfying $\alpha^4 = \alpha + 1$. For each of the 16 solutions to the linear recurrence in Example 9.3, find the value of θ such that $s_t = \mathrm{Tr}(\theta\alpha^t)$.

7. Let d and e be divisors of n. Show that

$$\gcd(\frac{n}{d}, \frac{n}{e}) = \frac{n}{de} \cdot \gcd(d, e).$$

Conclude from this that

$$N_1 = \gcd(N, \frac{q^m - 1}{q - 1}),$$

as claimed in (9.31).

8. Let $f(x)$ be irreducible and of period n over $GF(q)$. Show that $N_1 = 1$ if and only if $q \equiv 1 \pmod N$ and $\gcd(m, N) = 1$.

9. Given that the following polynomials are irreducible over the given fields, find n, N, d, e, N_1, and a set of N_1 sequences of length d which represent all possible solutions to the recursion with characteristic polynomial $f(x)$ over $GF(q)$ up to projective cyclic equivalence.

a. $f(x) = x^3 + x^2 + x + 2, \qquad q = 3.$

b. $f(x) = x^5 + 2x^3 + x^2 + 2x + 2, \qquad q = 3.$

c. $f(x) = x^2 + 3x + 1, \qquad q = 7.$

d. $f(x) = x^6 + x^5 + x^4 + x^2 + 1, \qquad q = 2.$

10. Prove Corollary 9.12.

11. Complete the details of the proof of Theorem 9.13.

12. Use Theorem 9.13 to give a formula for the Fibonacci numbers (mod 5).

13. This problem will explore the case of a characteristic polynomial with repeated roots a bit further than was done in the text.

a. Consider the recurrence over $GF(q)$ whose characteristic polynomial is $f(x) = (x-1)^m$, for some integer $m \geq 1$. Show that for any $0 \leq k \leq m-1$,

$$s_t = \binom{t}{k}, \qquad \text{for } t \geq 0$$

is a solution to this recursion.

b. Show that the m solutions found in part (a) are linearly independent.

c. Now consider a recurrence with characteristic polynomial $f(x) = (x-\alpha)^m$, for some $\alpha \in GF(q)$. Show that if (s_t) is any solution to the recurrence with characteristic polynomial $(x-1)^m$, then

$$g_t = s_t \alpha^t$$

is a solution to the recurrence with characteristic polynomial $(x-\alpha)^m$, and further that all solutions have this form.

d. Apply the results of parts (a), (b), and (c) to the Fibonacci recurrence mod 5. Give an explicit formula for the general solution, and specialize it to the particular case $s_0 = 0$ and $s_1 = 1$.

Chapter 10

The Theory of m-Sequences

In this chapter we will use the techniques developed in Chapter 9 to define and investigate the important class of binary sequences called *m-sequences.*[*] These sequences have fascinating mathematical properties and find important applications in telecommunications and computer science. Roughly speaking, they are important because they are easily-generated binary sequences that behave in many respects as if they were *completely random.*

An m-sequence is a binary sequence (not identically zero) that satisfies a linear recurrence whose characteristic polynomial is *primitive.* We remind you that a primitive polynomial (see Example 5.7) is an irreducible polynomial of degree m which is the minimal polynomial of a primitive root in $GF(2^m)$. Alternatively, a primitive polynomial is an irreducible factor of one of the cyclotomic polynomials Φ_{2^m-1}.

Since the characteristic polynomial of an m-sequence has period 2^m-1, it follows from Theorem 9.4 that every m-sequence has period $2^m - 1$.

Example 10.1. Using the five primitive polynomials x^2+x+1, x^3+x+1,

---

* These sequences have several other common names, e.g., *PN (pseudo-noise) sequences* and *maximal-length shift register sequences.*

x^4+x+1, and x^5+x^2+1 (x^5+x+1 is not primitive), and starting the sequence out with $0, 0, \ldots, 0, 1$, we find the following examples of m-sequences (we show just one period of each sequence).

m	$f(x)$	(s_t)
2	x^2+x+1	011
3	x^3+x+1	0010111
4	x^4+x+1	000100110101111
5	x^5+x^2+1	0000100101100111110001101110101

Each of these sequences can be produced by a simple linear feedback shift register. The shift registers needed to generate the m-sequence of length 7, for example, is shown in Figure 10.1. ■

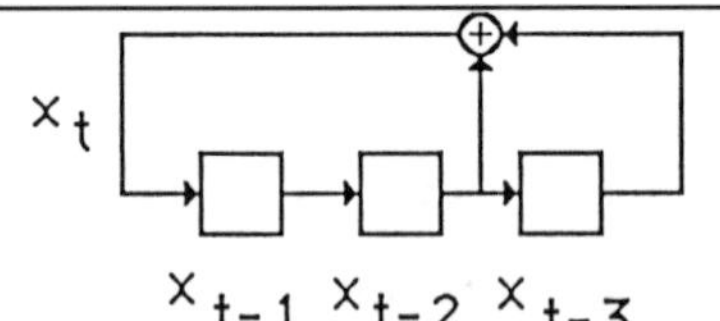

Figure 10.1. A circuit to implement the recursion $x_t = x_{t-2} + x_{t-3}$, whose characteristic polynomial is the primitive polynomial $x^3 + x + 1$.

The importance of m-sequences is due in large part to their so-called *pseudo randomness properties*, i.e., properties that make m-sequences behave like sequences whose elements are chosen at random. The next theorem gives the simplest yet most important of these properties. It refers to the m-grams of an m-sequence. If $(s_0, s_1, \ldots, s_{n-1})$ is the m-sequence in question, an m-*gram* is one of the n subsequences of length m of the form

$$(10.1) \qquad (s_t, s_{t+1}, \ldots, s_{t+m-1}), \qquad \text{for } t = 0, 1, \ldots, n-1,$$

where the subscripts in (10.1) are taken mod n if necessary. For example, the 3-grams of the m-sequence of length 7 in Example 10.1 are (001), (010),

(101), (011), (111), (110), and (100). Notice that they are all different and that every possible 3-gram except (000) appears. This is no accident!

Theorem 10.1. *Among the $2^m - 1$ m-grams of an m-sequence (s_t), each nonzero binary vector of length m occurs once and only once.*

Proof: All the m-grams are distinct, since a repeated m-gram would cause (s_t) to repeat sooner than period n, because of the degree m recurrence relation. On the other hand, the all-zero m-gram cannot occur, because if it did, the sequence (s_t) would continue to be 0 because of the degree m recurrence relation. ∎

One of the most remarkable of the pseudo-randomness properties of m-sequences is their *run-distribution properties*, which we now describe. A *run of length r* in a binary sequence is a subsequence of exactly r consecutive 1's (or 0's). The sequence 01101100 of length 8, for example, when viewed cyclically, has one 0-run of length 1, two 1-runs of length 2, and one 0-run of length 3.

It turns out that m-sequences have a very regular and predictable distribution of run lengths. Before stating a general theorem (Theorem 10.2, below), however, let us gather some data by counting the number of runs in the m-sequence of length 31 given in Example 10.1. A histogram of the runs follows (again, we view the sequence cyclically).

length	0-runs	1-runs
1	4	4
2	2	2
3	1	1
4	1	0
5	0	1
Totals:	8	8

There are a total of 16 runs; half have length 1, one-fourth have length 2, one-eighth have length 3. This is just what one would hope for in a completely random sequence 0's and 1's. (In fact it's too good to be true! In a random sequence of length 31, the average number of runs of various lengths is as given in the above table; but to get *exactly* the average for any particular sequence is extremely improbable.) This nice behavior happens for any m-sequence, and the general histogram for an m-sequence of length $2^m - 1$ is as follows.

length	0-runs	1-runs
1	2^{m-3}	2^{m-3}
2	2^{m-4}	2^{m-4}
$\vdots$	$\vdots$	$\vdots$
r	2^{m-r-2}	2^{m-r-2}
$\vdots$	$\vdots$	$\vdots$
$m-2$	1	1
$m-1$	1	0
m	0	1
Totals:	2^{m-2}	2^{m-2}

Theorem 10.2. *The run distribution for any m-sequence of length $2^m - 1$ is as given in the table above.*

Proof: Let $(s_0, s_1, \ldots, s_{n-1})$, $n = 2^m - 1$ denote the m-sequence in question. Using Theorem 10.1, we can count the runs of various lengths in an m-sequenceby distinguishing three cases:

• *Length m:* By Theorem 10.1, there are no 0-runs of length m, and just one 1-run of length m.

• *Length $m-1$:* By Theorem 10.1, we know that the m–gram 11111111 appears exactly once. Furthermore, it must be sandwiched between 0's; otherwise, there would be more than one all-ones m–gram. Thus the subsequence 0111111110 of length $m+2$ appears somewhere in the m-sequence. Within this subsequence appears the two m–grams 01111111 and 11111110. If there were a separate 1-run of length $m-1$ somewhere else in the m-sequence, it too would be sandwiched between 0's, leading to the subsequence 011111110, and this would lead to another appearance of the m–grams 01111111 and 11111110. Since by Theorem 10.1 these can occur only once each , there can be no separate 1–run of length $m-1$. There is, however, one 0–run of length $m-1$, because of the m–grams 10000000 and 00000001, which must, of course, occur together as 100000001.

• *Length $r \leq m-2$:* Each 1-run of length $r \leq m-2$ corresponds to an m–gram of the form

$$\overbrace{0\ \underbrace{1\,1\,1\,1\,1\,1\,1\,1}_{r}\ 0\underbrace{x\,x\,x\,x\,x}_{m-r-2}}^{m},$$

the x's denoting arbitrary binary digits. There are clearly 2^{m-r-2} such m–grams, and hence 2^{m-r-2} 1–runs of length r. A similar argument shows that there are 2^{m-r-2} 0–runs of length r. This completes the proof of Theorem 10.1. ∎

That concludes our short discussion of the run length properties of m-sequences. We next consider their important *correlation* properties.

Given two binary sequences $\mathbf{x} = (x_1, x_2, \ldots, x_n)$, $\mathbf{y} = (y_1, y_2, \ldots, y_n)$ of length n, their *correlation* $C(\mathbf{x}, \mathbf{y})$ is defined to be the number of *agreements* minus the number of *disagreements* between $\mathbf{x}$ and $\mathbf{y}$. That is,

$$C(\mathbf{x}, \mathbf{y}) = A - D,$$

where

$$A = |\{i : x_i = y_i\}| \qquad \text{and} \qquad D = |\{i : x_i \neq y_i\}|.$$

For example, if $\mathbf{x} = (00110011)$, $\mathbf{y} = (01010101)$, we have $A = |\{1,4,5,8\}| = 4$, $D = |\{2,3,6,7\}| = 4$, and so $C(\mathbf{x}, \mathbf{y}) = 0$. The correlation $C(\mathbf{x}, \mathbf{y})$ is a measure of the similarity between $\mathbf{x}$ and $\mathbf{y}$. If $\mathbf{x}$ and $\mathbf{y}$ are identical, then $C(\mathbf{x}, \mathbf{y}) = n$, and if $\mathbf{x}$ and $\mathbf{y}$ disagree in every component, then $C(\mathbf{x}, \mathbf{y}) = -n$. In every case, then, we have $-n \leq C(\mathbf{x}, \mathbf{y}) \leq n$ and very little more can be said in general, although the following lemma is sometimes useful.

Lemma 10.3. *Let $\mathbf{x}$ be a binary sequence of length n containing w_1 +1's, and let $\mathbf{y}$ be one containing w_2 +1's. Then*

$$C(\mathbf{x}, \mathbf{y}) \equiv n - 2(w_1 + w_2) \pmod 4.$$

Proof: Define

$$n_{1,1} = |\{i : x_i = 1,\ y_i = 1\}|$$
$$n_{0,1} = |\{i : x_i = 0,\ y_i = 1\}|$$
$$n_{1,0} = |\{i : x_i = 1,\ y_i = 0\}|$$
$$n_{0,0} = |\{i : x_i = 0,\ y_i = 0\}|.$$

Then we have $C(\mathbf{x}, \mathbf{y}) = A - D = (n_{1,1} + n_{0,0}) - (n_{0,1} + n_{1,0})$. But $n_{1,0} = w_1 - n_{1,1}$, $n_{0,1} = w_2 - n_{1,1}$, $n_{0,0} = n - w_1 - w_2 + n_{1,1}$, as illustrated in the following diagram:

$$\overbrace{\begin{matrix}1\,1\,1\,1\,1\,1\,1\\1\,1\,1\,1\,1\,1\,1\end{matrix}}^{n_{1,1}}\quad \overbrace{\begin{matrix}1\,1\,1\,1\,1\\0\,0\,0\,0\,0\end{matrix}}^{w_1 - n_{1,1}}\quad \overbrace{\begin{matrix}0\,0\,0\,0\,0\,0\\1\,1\,1\,1\,1\,1\end{matrix}}^{w_2 - n_{1,1}}\quad \overbrace{\begin{matrix}0\,0\,0\,0\,0\,0\,0\\0\,0\,0\,0\,0\,0\,0\end{matrix}}^{n - w_1 - w_2 + n_{1,1}}$$

Hence

$$\begin{aligned} C(\mathbf{x}, \mathbf{y}) &= n - 2w_1 - 2w_2 + 4n_{1,1} \\ &\equiv n - 2(w_1 + w_2) \pmod{4}. \end{aligned}$$ ∎

Given a fixed sequence $\mathbf{x}$, its *autocorrelation function* $C(\tau)$ is defined to be the correlation between $\mathbf{x}$ and its τth cyclic shift. For computational purposes, it is best to assume that the components of $\mathbf{x}$ are $+1$, -1, rather than 0, 1. With this convention $C(\mathbf{x}, \mathbf{y}) = \sum x_i y_i$, and the autocorrelation function $C(\tau)$ can be written as

$$C(\tau) = \sum_{i=1}^{n} x_i x_{i+\tau} \qquad \text{(subscripts mod } n\text{)}. \tag{10.2}$$

If the components are instead 0's and 1's, the formula (10.2) becomes

$$C(\tau) = \sum_{i=1}^{n} (-1)^{x_i + x_{i+\tau}}, \tag{10.3}$$

where the addition in the exponent is mod 2 addition. The following lemma now follows immediately from Lemma 10.3.

Corollary 10.4. *For any binary sequence, the autocorrelation function satisfies*

$$C(\tau) \equiv n \pmod 4.$$

Proof: Take $w_1 = w_2$ in Lemma 10.3. ∎

Theorem 10.1 says that as far as *runs* are concerned, m-sequences behave very much like a sequence generated at random by flipping a coin. The next theorem, Theorem 10.4, says that in terms of autocorrelation, m-sequences also behave much like random sequences. To see that this is the case, let's briefly consider the autocorrelation function of a truly random sequence. Thus let

$$\mathbf{X} = (X_1, X_2, \ldots, X_n)$$

be a sequence of n independent, identically distributed random variables with distribution function $\Pr\{X = +1\} = \Pr\{X = -1\} = 1/2$. The values assumed by the autocorrelation $C(\tau)$ function of $\mathbf{X}$ are themselves random variables:

$$C(\tau) = \sum_{i=1}^{n} X_i X_{i+\tau} \qquad \text{(subscripts mod } n\text{)}.$$

Let us denote the expectation of $C(\tau)$ by $\widehat{C}(\tau)$. Then

$$\begin{aligned} \hat{C}(\tau) &= \sum_{i=1}^{n} E(X_i X_{i+\tau}) \\ &= \begin{cases} 0 & \text{if } \tau \not\equiv 0 \pmod n \\ n & \text{if } \tau \equiv 0 \pmod n, \end{cases} \end{aligned}$$

because $E(X_iX_{i+\tau}) = E(X_i)E(X_{i+\tau}) = 0 \cdot 0 = 0$, if $\tau \not\equiv 0 \pmod{n}$, and $E(X_iX_{i+\tau}) = E(X_i^2) = E(1) = 1$, if $\tau \equiv 0 \pmod{n}$. We conclude that for $\tau \not\equiv 0 \pmod{n}$, a random sequence should have very small values of $C(\tau)$, on the average. The next theorem shows that m-sequences have this property also.

Theorem 10.5. *If $(s_t)_{t=0}^{n-1}$ is an m–sequence of length $n = 2^m - 1$, then*

$$\begin{aligned} C(\tau) &= -1 \qquad \text{if } \tau \not\equiv 0 \pmod{n} \\ &= \ \ n \qquad \text{if } \tau \equiv 0 \pmod{n}. \end{aligned}$$

Proof: The second assertion simply says that the m-sequence agrees with itself in every position. To prove the first, we use Theorem 9.2, which implies that if (s_t) is an m-sequence,

$$s_t = \mathrm{Tr}(\theta\alpha^t), \tag{10.4}$$

where α is a primitive root in $GF(2^m)$ and θ is a nonzero element of $GF(2^m)$. Thus also

$$s_{t+\tau} = \mathrm{Tr}(\theta\alpha^{t+\tau}),$$

and so by (10.3),

$$\begin{aligned} C(\tau) &= \sum_{t=0}^{n-1} (-1)^{\mathrm{Tr}(\theta\alpha^t)+\mathrm{Tr}(\theta\alpha^{t+\tau})} \\ &= \sum_{t=0}^{n-1} (-1)^{\mathrm{Tr}(\theta\alpha^t(1+\alpha^\tau))}. \end{aligned}$$

If $\tau \not\equiv 0 \pmod{n}$, then $1+\alpha^\tau \neq 0$, and as t runs from 0 to $n-1$, $(1+\alpha^\tau)\alpha^t$ runs through the nonzero elements of $GF(2^m)$. By Theorem 8.1, this means that $\mathrm{Tr}((1+\alpha^\tau)\alpha^t)$ is one 2^{m-1} times and zero $2^{m-1}-1$ times. Hence the above sum for $C(\tau)$ is $(2^{m-1}-1)-(2^{m-1}) = -1$. ∎

There are many other binary sequences that share with m-sequences the property of Theorem 10.4, for example, the sequence 01011100010 of length 11, or the sequence 0100111101010000110 of length 19. But only m-sequences have the stronger *cycle-and-add property* described in Theorem 10.6.

Theorem 10.6. *Let $(s_t)_{t=0}^{n}$ be an m-sequence of length $n = 2^m - 1$. Then for any $\tau \not\equiv 0 \pmod{n}$, there exists an unique integer σ, with $1 \leq \sigma \leq n-1$, such that*

$$s_t + s_{t+\tau} = s_{t+\sigma} \qquad \textit{for all } t \geq 0.$$

Proof: As we observed above (see (10.4)) Theorem 9.2 implies that there exists a primitive root $\alpha \in GF(2^m)$ such that

$$s_t = \mathrm{Tr}(\theta\alpha^t),$$

for some nonzero element θ of $GF(2^m)$. But since α is a primitive root, θ must be a power of α, say $\theta = \alpha^i$. Thus

$$s_t = \mathrm{Tr}(\alpha^{t+i}) \qquad \text{for } t \geq 0.$$

Hence

$$s_{t+\tau} = \mathrm{Tr}(\alpha^{t+\tau+i}),$$

and so

$$\begin{aligned} s_t + s_{t+\tau} &= \mathrm{Tr}(\alpha^{t+i}) + \mathrm{Tr}(\alpha^{t+\tau+i}) \\ &= \mathrm{Tr}(\alpha^{t+i}(1+\alpha^{\tau})). \end{aligned} \tag{10.5}$$

Thus if $1 + \alpha^{\tau} \neq 0$, i.e., if $\tau \not\equiv 0 \pmod{n}$, then $\alpha^{\tau} + 1 = \alpha^{\sigma}$ for some σ, $1 \leq \sigma \leq n-1$, and (10.5) becomes simply

$$\begin{aligned} s_t + s_{s+\tau} &= \mathrm{Tr}(\alpha^{t+i+\sigma}) \\ &= s_{t+\sigma}, \end{aligned}$$

as asserted. ∎

Example 10.2. Let $(s_t) = 0010111\cdots$ be an m-sequence of period 7 (cf. Example 10.1). Let us add s_t and s_{t+4}:

$$\begin{array}{rcl} s_t & = & 0\,0\,1\,0\,1\,1\,1 \quad \cdots \\ s_{t+4} & = & 1\,1\,1\,0\,0\,1\,0 \quad \cdots \\ \hline & & 1\,1\,0\,0\,1\,0\,1 \quad \cdots \end{array}$$

and by inspection, $s_t + s_{t+4} = s_{t+5}$. ∎

A given m-sequence has $2^m - 1$ different cyclic shifts, but there is a "canonical" one which is often useful. It is described in the next theorem.

Theorem 10.7. *Let (s_t) be an m-sequence. Then there exists an integer τ (unique, mod n) such that the shifted sequence $r_t = s_{t+\tau}$ satisfies*

$$r_t = r_{2t} \qquad \text{for all } t \geq 0. \tag{10.6}$$

Proof: As we have seen above above, the given m-sequence will be of the form $s_t = \mathrm{Tr}(\alpha^{t+i})$ for some integer i. Choose $\tau = n - i$. Then the sequence $r_t = s_{t+\tau}$ is given by

$$r_t = \mathrm{Tr}(\alpha^{t+i+n-i})$$

$$r_t = \mathrm{Tr}(\alpha^t). \tag{10.7}$$

Thus $r_{2t} = \mathrm{Tr}(\alpha^{2t})$. But since $\mathrm{Tr}(x) = \mathrm{Tr}(x^2)$ for all $x \in GF(2^m)$, it follows that $r_{2t} = \mathrm{Tr}(\alpha^t) = r_t$.

To show that τ is unique, assume that the sequence $u_t = \mathrm{Tr}(\alpha^{t+i})$ satisfies (10.6), i.e.,

$$\mathrm{Tr}(\alpha^{2t}\alpha^i) = \mathrm{Tr}(\alpha^{t+i}) \qquad \text{for } t \geq 0.$$

Then because

$$\mathrm{Tr}(\alpha^{t+i}) = \mathrm{Tr}(\alpha^{2t+2i}),$$

it follows that

$$\operatorname{Tr}(\alpha^{2t+i}) = \operatorname{Tr}(\alpha^{2t+2i})$$

i.e.,

$$\operatorname{Tr}(\alpha^{2t}(\alpha^i + \alpha^{2i})) = 0, \qquad \text{for } t \geq 0.$$

Thus by Lemma 9.3, $\alpha^i + \alpha^{2i} = 0$, i.e., $\alpha^i = 1$, or $i = 0$. Thus $u_t = \operatorname{Tr}(\alpha^t)$ is the unique solution to (10.6). ∎

The next question we wish to ask is "How many different m-sequences of length $n = 2^m - 1$ are there?" Let us agree that two m-sequences which are translates of each other are the same. This means that two m-sequences which are produced by the same linear recursion are the same, and so the number of m-sequences is equal to the number of primitive polynomials of degree m. But by Theorem 7.2, this number is $\phi(2^m - 1)/m$. We have therefore proved

Theorem 10.8. *There are exactly $\phi(2^m - 1)/m$ (cyclically) distinct m-sequences of length $2^m - 1$.*

Here is a short table of $\phi(2^m - 1)/m$:

m	$\phi(2^m-1)/m$
1	1
2	1
3	2
4	2
5	6
6	6
7	18
8	16
9	48
10	60
⋮	⋮

It turns out that if we are given just one m-sequence of length $2^m - 1$, it is relatively easy to find all the others. For example, let $m = 4$; in

Example 10.1 we saw that 000100110101111 is one m-sequenceof length 15. According to Theorem 10.8, there is another one, one that is not just a cyclic shift of the given one. How could we find it? One way is to use x^4+x+1 to build the field $GF(16)$, compute minimal polynomials, and discover that x^4+x^3+1 is the other primitive polynomial of degree 4, and use it to construct the other m-sequence. There is a much better way, however, which involves the notion of *decimation.*

Given a sequence $(s_t)_{t=0}^{\infty}$, and any integer $d \geq 1$, a dth *decimation* of (s_t) is any sequence (r_t) obtained by taking every dth term of the original sequence. If we start with the element s_i of the original sequence, then the new sequence r_t is given by

$$r_t = s_{td+i} \qquad t \geq 0. \tag{10.8}$$

Note that the m-sequences described in Theorem 10.7 are *invariant* under decimation by 2, if we start at s_0. The next theorem and its corollary show that any m-sequence of length $2^m - 1$ can be obtained from any other by decimation.

Theorem 10.9. *If (s_t) and (r_t) are two m-sequences of length 2^m-1, which satisfy (10.6), then there exists an integer d, relatively prime to $2^m - 1$, such that*

$$r_t = s_{td} \qquad \text{for all } t \geq 0.$$

Proof: By Theorem 10.7 (s_t) and (r_t) can be written as follows:

$$\begin{aligned} s_t &= \mathrm{Tr}(\alpha^t), & \mathrm{ord}(\alpha) &= 2^m - 1 \\ r_t &= \mathrm{Tr}(\beta^t), & \mathrm{ord}(\beta) &= 2^m - 1. \end{aligned}$$

But by Lemma 5.4 and the remarks immediately preceding it, there exists $1 \leq d \leq n-1$ with $\gcd(n,d) = 1$ such that $\beta = \alpha^d$. Thus

$$r_t = \mathrm{Tr}(\alpha^{dt}) = s_{td}. \qquad \blacksquare$$

Corollary 10.10. *Any m-sequence of length $2^m - 1$ can be obtained from any other by suitable decimation.*

Proof: Let (s'_t) and (r'_t) be any two m-sequences of length $2^m - 1$, Then by Theorem 10.6, we have, for certain integers i and j,

$$\begin{aligned} s'_{t+i} &= \mathrm{Tr}(\alpha^t) \\ r'_{t+j} &= \mathrm{Tr}(\beta^t), \end{aligned}$$

where the sequences $(s_t) = (s'_{t+i})$ and $(r_t) = (r'_{t+j})$ satisfy the condition (10.6). Thus by Theorem 10.9, $r_t = s_{td}$ for some integer d. It follows that

$$\begin{aligned} r'_{t+j} &= s'_{(t+i)d}, \qquad \text{i.e.,} \\ r'_t &= s'_{(t+i-j)d} = s'_{td+(i-j)d}. \end{aligned}$$

Therefore (r_t) can be obtained by decimating (s_t) by d, starting with the term $s_{(i-j)d}$. ∎

Example 10.3. Consider the m-sequence of length 7 $s_t = \mathrm{Tr}(\alpha^t)$ where α is a root of the primitive polynomial $f(x) = x^3 + x + 1$. Then (e.g. by Theorem 8.3), $\mathrm{Tr}(1) = 1$, $\mathrm{Tr}(\alpha) = 0$, $\mathrm{Tr}(\alpha^2) = 0$, and so

$$(s_t) = 1\,0\,0\,1\,0\,1\,1\,1\,0\,0\,1\,0\,1\,1\,\cdots$$

If we decimate (s_t) by $d = 2$, i.e. take every second term, we get the sequence

$$(r_t) = 1\,0\,0\,1\,0\,1\,1\,\cdots,$$

which is just (s_t) back again. This is exactly what Theorem 10.7 predicts, since $r_t = \mathrm{Tr}(\alpha^{2t}) = \mathrm{Tr}(\alpha^t) = s_t$. Similarly if we decimate by $d = 4$, 8, 16, etc., we continue to get (s_t). But taking $d = 3$, we get a new sequence:

$$(r_t) = (s_{3t}) = 1\,1\,1\,0\,1\,0\,0\,\cdots,$$

which is a different m-sequence, this time with characteristic polynomial $(x - \alpha^3)(x - \alpha^6)(x - \alpha^{12}) = x^4 + x^3 + 1$. Similarly, $d = 5$ or $d = 6$ gives this same m-sequence, since α^5 and α^6 are conjugates of α^3. ∎

The general situation is this. Given an m-sequence of length n, we can by Theorem 10.9 decimate it by any number d lying in the set

$$\Gamma = \{1 \le d \le n : \gcd(d, n) = 1\}$$

and get another m-sequence. This new m-sequence will be the same as the original if and only if d lies in the subset

$$\Gamma_0 = \{d = 2^j : j = 0, 1, \ldots, m-1\}.$$

Actually, Γ is best viewed as a multiplicative group (do the multiplication mod n), and Γ_0 as a subgroup. Then two values of d will yield the same m-sequence if and only if they lie in the same coset of Γ_0.

For example, if $m = 6$, the group Γ has $\phi(63) = 36$ elements, and the subgroup Γ_0 is

$$\Gamma_0 = \{1, 2, 4, 8, 16, 32\}.$$

The cosets of Γ_0 with respect to Γ are:

$$\begin{aligned}
\Gamma_0 &= \{\ 1,\ 2,\ 4,\ 8, 16, 32\} \\
5\Gamma_0 &= \{\ 5, 10, 20, 40, 17, 34\} \\
11\Gamma_0 &= \{11, 22, 44, 25, 50, 37\} \\
13\Gamma_0 &= \{13, 26, 52, 41, 19, 38\} \\
23\Gamma_0 &= \{23, 46, 29, 58, 53, 43\} \\
32\Gamma_0 &= \{31, 62, 61, 59, 55, 47\}.
\end{aligned}$$

These cosets together account for all 36 elements in Γ; and to get all 6 cyclically distinct m-sequences of length 63, given just one such, we decimate it by (e.g.) 5, 11, 13, 23, and 31.

Problems for Chapter 10.

1. The polynomial $f(x) = x^5 + x + 1$ was not used in Example 10.1 because it is not primitive.

a. If $f(x)$ is used as the characteristic polynomial for a linear recurring sequence over $GF(2)$, beginning 00001, what is the resulting period?
b. Based on the results of part (a), what can you say about the factorization of $f(x)$?
c. Use Berlekamp's algorithm (Chapter 7) to factor $f(x)$ completely.

2. Draw shift registers to implement the m-sequences in Example 10.1 with characteristic polynomials x^2+x+1, x^4+x+1, and x^5+x+1.

3. Notice that the shift register in Figure 10.1 is identical to the one in Figure 8.2. Is this an accident, or an instance of a general law?

4. Show that if 2^m-1 is prime, then any irreducible polynomial of degree m over $GF(2)$ is also primitive. Give an example that shows that this is false when 2^m-1 is not prime.

5. Suppose a binary sequence of length 31 is selected at random. Calculate the expected number of 0-runs and 1-runs of lengths 1, 2, 3, 4, and 5. Compare your result to Theorem 10.2.

6. According to Theorem 10.1, each of the 31 different nonzero 5-grams appears once in the m-sequence of length 31 given in Example 10.1. In this problem, we ask you to consider the *6-grams* of that sequence.
a. How many of the 64 possible 6-grams appear in the length 31 m-sequence of Example 10.1?
b. Show that there exists a fixed length 6 vector $\mathbf{a}$ with the property that the 6-gram $\mathbf{x}$ appears in the m-sequence if and only if $\mathbf{x}$ is not zero and $\mathbf{x}\cdot\mathbf{a}=0$. Find $\mathbf{a}$ explicitly.

7. Verify that the two sequences 01011100010 and 0100111101010000110 have $C(\tau)$'s satisfying the conclusions of Theorem 10.5.

8. Use the m-sequence of length 15 from Example 10.1, and add (s_t) and (s_{t+4}). Which shift results?

9. In Example 10.2 we added the m-sequences (s_t) and (s_{t+4}), and found that the resulting sequence was (s_{t+5}). For each sequence of the form (s_{t+i}) for $i = 1, 2, 3, 5$, and 6, add (s_t) and (s_{t+i}) and identify the resulting sequence as (s_{t+j}) for some j.

10. For each of the four m-sequences listed in Example 10.1, find the cyclic shift of it that satisfies $s_t = s_{2t}$. (Cf. Theorem 10.7.)

11. The polynomial $f(x) = x^6 + x + 1$ is primitive.

a. Find the m-sequence with characteristic polynomial $f(x)$ which satisfies $s_t = s_{2t}$.

b. Find all other m-sequences of length 63 which satisfy $s_t = s_{2t}$.

12. Let $(s_t) = 0010111\cdots$ be the m-sequence of period 7 given in Example 10.1. For each value of τ from 1 to 6, find the corresponding value of σ such that

$$s_t + s_{t+\tau} = s_{t+\sigma} \qquad \text{for all } t \geq 0.$$

13. Prove the following two facts about $C(\mathbf{x}, \mathbf{y})$.

a. $C(\mathbf{x}, \mathbf{y})$ must have the same parity as n.

b. For every number c between $+n$ and $-n$ such that $c \equiv n \pmod 2$, there exist sequences $\mathbf{x}$ and $\mathbf{y}$ such that $C(\mathbf{x}, \mathbf{y}) = c$.

14. Any m-sequence can be used to construct a *continuous function of time* with some interesting properties. In this problem we will investigate this construction. If (s_n) is an m-sequence with period $2^m - 1$, define the real-valued function $S(t)$ as follows:

$$S(mt_0) = \begin{cases} +1, & \text{if } s_m = 0; \\ -1, & \text{if } s_m = 1, \end{cases}$$

where t_0 is a fixed positive number. For values t which are not integer multiples of t_0, $S(t)$ is defined by linear interpolation.

a. What is the period of the function $S(t)$?

b. The *autocorrelation function* of the function $S(t)$ is defined to be

$$C(\tau) = \int_0^{T_0} S(t)S(t+\tau)dt,$$

where T_0 is the period of $S(t)$ found in part (a). Find $C(\tau)$ explicitly. Compare your result to Theorem 10.5.

Chapter 11

Crosscorrelation Properties of m-Sequences

In the last chapter we defined m-sequences and studied some of their elementary properties. In this chapter we will continue our study of m-sequences, and in particular we will study the *crosscorrelation* properties of m-sequences. Before doing so, indeed even before defining the crosscorrelation function, we pause to explain one reason why these crosscorrelation properties are so important.

Consider the following simple model for a communication system. There are two users, called A and B. Each wishes to transmit a sequence of data bits (± 1) to a receiver C at a distant destination. We denote A's data bits by $(a_1, a_2, a_3, \ldots)$, each a_i being ± 1; B's data bits are similarly $(b_1, b_2, \ldots)$.

The two users are each assigned a *signature sequence*, i.e., a binary (± 1) sequence of a fixed length n. A is assigned the sequence $\mathbf{x} = (x_0, x_1, \ldots, x_{n-1})$, and B the sequence $\mathbf{y} = (y_0, y_1, \ldots, y_{n-1})$. We suppose that A uses the signature sequence $\mathbf{x}$ as a "carrier," and transmits to C the sequence of *vectors*

$$T_A = (a_1\mathbf{x}, a_2\mathbf{x}, a_3\mathbf{x}, \ldots),$$

while B transmits

$$T_B = (b_1\mathbf{y}, b_2\mathbf{y}, b_3\mathbf{y}, \ldots).$$

We assume that C receives not T_A and T_B separately, but rather the *sum*

$$R_C = (\mathbf{z}_1, \mathbf{z}_2, \ldots), \qquad \text{where } \mathbf{z}_i = a_i\mathbf{x} + b_i\mathbf{y}.$$

Here is the problem. How can C recover A's message from R_C? The best way is for C to estimate A's ith transmitted bit a_i as

$$\hat{a}_i = \tfrac{1}{n}(\mathbf{x} \cdot \mathbf{z}_i), \tag{11.1}$$

where $(\mathbf{x} \cdot \mathbf{z}_i)$ denote the scalar ("dot") product of the two vectors. A simple calculation based on (11.1) leads to the error estimate

$$|\hat{a}_i - a_i| = \tfrac{1}{n}(\mathbf{x} \cdot \mathbf{y}). \tag{11.2}$$

The discrepancy between a_i and $\hat{a}_i$ given in (11.2) is often called "crosstalk." It is entirely due to the non-orthogonality of the two signature sequences $\mathbf{x}$ and $\mathbf{y}$. If this were all there was to the story, crosstalk could be eliminated by simply choosing $\mathbf{x}$ and $\mathbf{y}$ to be orthogonal. But a complication arises.

The complication is that the transmitted streams T_A and T_B may be *out of phase.* Thus if say $n = 5$, $(a_i) = (+1, -1, -1, +1, \cdots)$, $(b_i) = (+1, +1, +1, -1, \cdots)$, then T_A and T_B might look like this:

$$T_A = x_0\, x_1\, x_2\, x_3\, x_4\, \overline{x_0}\, \overline{x_1}\, \overline{x_2}\, \overline{x_3}\, \overline{x_4}\, \overline{x_0}\, \overline{x_1}\, \overline{x_2}\, \overline{x_3}\, \overline{x_4}\, x_0\, x_1\, x_2\, x_3\, x_4 \ldots$$

$$T_B = y_2\, y_3\, y_4\, y_0\, y_1\, y_2\, y_3\, y_4\, y_0\, y_1\, y_2\, y_3\, y_4\, \overline{y_0}\, \overline{y_1}\, \overline{y_2}\, \overline{y_3}\, \overline{y_4} \cdots$$

where overbars denote negation. Now if C attempted to estimate a_1 using formula (11.1), he or she would obtain

$$\hat{a}_1 = \tfrac{1}{5}(5 + x_0y_2 + x_1y_3 + x_2y_4 + x_3y_0 + x_4y_1),$$

with a corresponding error of

$$|\hat{a}_1 - a_1| = \tfrac{1}{5}(x_0y_2 + x_1y_3 + x_2y_4 + x_3y_0 + x_4y_1).$$

More generally, if the sequence T_B were shifted τ units with respect to the sequence T_A, and (11.1) was used to estimate $\hat{a}_i$, the error would be

$$|\hat{a}_i - a_i| = \tfrac{1}{n}(\mathbf{x} \cdot S^\tau \mathbf{y}), \tag{11.3}$$

where the *cyclic shift operator* S^τ is defined by

$$S^1\mathbf{y} = (y_{n-1}, y_0, \ldots, y_{n-2})$$

and

$$S^\tau \mathbf{y} = S(S^{\tau-1}\mathbf{y}).$$

This leads us to define the *crosscorrelation function* between $\mathbf{x}$ and $\mathbf{y}$, as follows:

$$\begin{aligned} C(\tau) &= \mathbf{x} \cdot S^\tau \mathbf{y} \\ &= \sum_{i=0}^{n-1} x_i y_{i+\tau}, \end{aligned} \tag{11.4}$$

where again subscripts are taken mod n. With this definition, the error between the true value of the data bit a_i and the $\hat{a}_i$ estimate (11.1) is

$$|\hat{a}_i - a_i| = \tfrac{1}{n}C(0)$$

if the a and b sequences are in phase, and

$$|\hat{a}_i - a_i| = \tfrac{1}{n}C(\tau), \tag{11.5}$$

if they are τ units out of phase. Since in real communications systems there is often no way to control the phase difference between the transmitters, there is considerable interest in finding pairs of binary sequences for which the crosscorrelation function is small for all values of τ. But even (11.5) does not tell the whole story, as can be seen by attempting to estimate a_3 using the example above. Here we obtain

$$\begin{aligned} \hat{a}_3 &= \tfrac{1}{5}(-5 + x_0y_2 + x_1y_3 + x_2y_4 - x_3y_0 - x_4y_1), \\ |\hat{a}_3 - a_3| &= \tfrac{1}{5}|(-x_0y_2 - x_1y_3 - x_2y_4 + x_3y_0 + x_4y_1)|. \end{aligned}$$

This error term is not a value assumed by the crosscorrelation function, because of the difference in the signs of the data bits. In general, we see that (11.5) might not be true; instead we might have

$$|\hat{a}_i - a_i| = \tfrac{1}{n}|\mathbf{x} \cdot N^\tau \mathbf{y}|, \tag{11.6}$$

where in (11.6) the *negacyclic shift* operator N is defined by

$$N^1\mathbf{y} = (-y_{n-1}, y_0, \ldots, y_{n-2}),$$

and

$$N^\tau\mathbf{y} = N^1(N^{\tau-1}\mathbf{y}).$$

The quantity $\mathbf{x} \cdot N^\tau\mathbf{y}$, when viewed as a function of the parameter τ, is sometimes called the *odd* crosscorrelation function between $\mathbf{x}$ and $\mathbf{y}$, and denoted by $\overline{C}(\tau)$:

$$\overline{C}(\tau) = \mathbf{x} \cdot N^\tau\mathbf{y}.$$

Unfortunately, it is very difficult to obtain analytic information about $\overline{C}(\tau)$. In practice, signature sequences are chosen in a two-step process. First, pairs of sequences, usually m-sequences, with uniformly low $C(\tau)$'s are selected using analytic techniques (some of which we develop below). Second, the values of $\overline{C}(\tau)$ are computed numerically for these pairs. And those pairs for which both $C(\tau)$ and $\overline{C}(\tau)$ are uniformly low are used in communication systems. At any rate, we hope this discussion makes you want to read what we have to say about $C(\tau)$!

So here is the definition we will use for the rest of the chapter. Given two binary sequences $\mathbf{x} = (x_0, \ldots, x_{n-1})$ and $\mathbf{y} = (y_0, \ldots, y_{n-1})$ (components: ± 1) of length n, we define their *crosscorrelation function*, $C(\tau)$, by

$$C(\tau) = \sum_{i=0}^{n-1} x_i y_{i+\tau}, \tag{11.7}$$

with subscripts taken mod n. In the special case that $\mathbf{x}$ and $\mathbf{y}$ are both m-sequences, we shall find that the calculation of $C(\tau)$ leads to some very interesting algebraic questions, and to some quite low values of $C(\tau)$.

Example 11.1. Let us compute the crosscorrelation function for the two cyclically inequivalent m-sequences of length 7, namely

$$(-1,-1,-1,+1,-1,+1,+1), \qquad \textit{and} \qquad (-1,+1,+1,-1,+1,-1,-1).$$

Calling these two sequences $\mathbf{x}$ and $\mathbf{y}$, respectively, we find by direct calculation that $C(0) = \mathbf{x} \cdot \mathbf{y} = -5$, $C(1) = \mathbf{x} \cdot S\mathbf{y} = +3$, etc. The complete table of the values for $C(\tau)$ is as follows.

τ	$C(\tau)$
0	−5
1	3
2	3
3	−1
4	3
5	−1
6	−1

Thus we find that $|C(\tau)| \leq 5$, for all values of τ. We shall see below a way of reaching this conclusion without ever having to directly compute a single value of $C(\tau)$! ■

For the remainder of this chapter, we will study the crosscorrelation function $C(\tau)$ for pairs of m-sequences. Thus let (x_t) and (y_t) be two m-sequences. By Theorem 10.7 we know that each of these sequences can be shifted so that it is of the form $(\mathrm{Tr}(\gamma^t))$ for some primitive root $\gamma \in GF(2^m)$. Since shifting (x_t) or (y_t) (or both) will not change the set of values assumed by the crosscorrelation function $C(\tau)$, from now on we will assume that in fact both sequences have this form. Further, Theorem 10.9 allows us to assume in fact that

$$\begin{aligned} x_t &= \mathrm{Tr}(\alpha^t) \\ y_t &= \mathrm{Tr}(\alpha^{dt}), \end{aligned}$$

where α is a primitive root in $GF(2^m)$ and d is some integer in the range $\{1, 2, \ldots, n-1\}$ which is relatively prime to n. Then by the definition (11.7), the crosscorrelation function for these two m-sequences is given by

$$C(\tau) = \sum_{t=0}^{n-1} (-1)^{\mathrm{Tr}(\alpha^{t-\tau} + \alpha^{dt})}. \tag{11.8}$$

If in the sum (11.8) we denote the constant $\alpha^{-\tau}$ by β, and note that as t runs from 0 to $n-1$, α^t runs through the nonzero elements of $GF(2^m)$, we obtain the alternative formula

$$C(\tau) = \sum_{x \neq 0} (-1)^{Tr(\beta x + x^d)}, \qquad \text{where } \beta = \alpha^{-\tau}, \tag{11.9}$$

the sum running over all nonzero elements in $GF(2^m)$. Sums of the form (11.9) are special cases of what number theorists call *exponential* sums and some of them have been extensively studied. For example, if $d = -1$, the sum in (11.9) becomes

$$C(\tau) = \sum_{x \neq 0} (-1)^{Tr(\beta x + x^{-1})}, \tag{11.10}$$

which is known as a *Kloosterman sum*. It is known (but is extremely difficult to prove!) that this sum satisfies $|C(\tau)| \leq 2^{m/2+1}$. For example, with $m = 3$, we get $|C(\tau)| \leq 4\sqrt{2} = 5.657\cdots$. But since $C(\tau)$ must be an integer, we in fact have $|C(\tau)| \leq 5$. Also, since by Lemma 10.3, $C(\tau) \equiv -1 \pmod 4$ it follows that the only possible values for $C(\tau)$ are -5, -1, $+3$. And we saw in Example 11.1 that in fact all three of these values are taken on by $C(\tau)$. (It is an open problem whether every value C such that $|C| \leq 2^{m/2+1}$ and $C \equiv -1 \pmod 4$ is assumed by the crosscorrelation function between an m-sequence and its $d = -1$ decimation, i.e., its reverse.)

Let us now consider a special case, a very important one, of the $C(\tau)$ problem that we will be able to handle with more-or-less elementary techniques. This case is $d = 2^e + 1$. However, decimating an m-sequence by $2^e + 1$ will produce another m-sequence if and only if $\gcd(2^e + 1, 2^m - 1) = 1$. The following lemma will identify the e's with this property.

Lemma 11.1. *For $1 \leq e \leq m$,*

$$\gcd(2^e+1, 2^m-1) = \begin{cases} 1, & \text{if } \gcd(2e,m) = \gcd(e,m) \\ 2^{\gcd(e,m)}+1, & \text{if } \gcd(2e,m) = 2\gcd(e,m). \end{cases}$$

Proof: Since $(2^e+1)(2^e-1) = 2^{2e}-1$, we have

$$\gcd(2^e+1, 2^m-1) \mid \gcd(2^{2e}-1, 2^m-1) = 2^{\gcd(2e,m)}-1,$$

by Theorem 2.3. If $\gcd(2e,m) = \gcd(e,m)$, this implies that

$$\gcd(2^e+1, 2^m-1) \mid 2^{\gcd(e,m)}-1 \mid 2^e-1.$$

But since 2^e+1 and 2^e-1 are relatively prime, this means that $\gcd(2^e+1, 2^m-1) = 1$ in this case. If the second alternative holds, i.e. if $\gcd(2e,m) = 2\gcd(e,m)$, then we have instead

$$\gcd(2^e+1, 2^m-1) \mid (2^{\gcd(e,m)}+1)(2^{\gcd(e,m)}-1).$$

But since 2^e+1 and $2^{\gcd(e,m)}-1$ are relatively prime, it follows that

$$\gcd(2^e+1, 2^m-1) \mid 2^{\gcd(e,m)}+1.$$

But

$$2^{\gcd(e,m)}+1 \mid 2^{\gcd(2e,m)}-1 \mid 2^m-1,$$

and

$$2^{\gcd(e,m)}+1 \mid 2^e+1,$$

since $e/\gcd(e,m)$ is odd. Thus

$$\gcd(2^e+1, 2^m-1) = 2^{\gcd(e,m)}+1$$

in this case, as asserted. ∎

It turns out that we will be able to evaluate the sum (11.9) explicitly whenever d is one more than a power of two, whether or not the gcd described in Lemma 11.1 is one, and so for the rest of this chapter, when we refer to $C(\tau)$, we will mean the sum in (11.9) rather than the crosscorrelation function between two m-sequences. It should be borne in mind, however, that only in the case $\gcd(2e, m) = \gcd(e, m)$ will our results imply anything about the crosscorrelation function between m-sequences. (The more general results will be applied at the end of the chapter when we discuss *Gold sequences.*)

Before launching into the general problem of computing the sums (11.9) when $d = 2^e + 1$, we will consider a specific numerical example at some length. It will lead us to the same result as Example 11.1, by a very different route!

Example 11.2. Let $d = 3$, $m = 3$. The function in the exponent of (11.9) is then $\mathrm{Tr}(\beta x + x^3) = \mathrm{Tr}(\beta x) + \mathrm{Tr}(x^3)$. The term $\mathrm{Tr}(\beta x)$ is a fairly simple linear function of x. But the term $\mathrm{Tr}(x^3)$ is more complicated. Let's examine it in some detail.

Let us write the function $\mathrm{Tr}(x^3)$ as a function of the *components* of the *vector* x. Thus let $x = x_0 + x_1\alpha + x_2\alpha^2$ with $x_i = 0$ or 1, where α is a primitive root satisfying $\alpha^3 = \alpha + 1$ in $GF(8)$. Since $\mathrm{Tr}(1) = 1$, $\mathrm{Tr}(\alpha) = \mathrm{Tr}(\alpha^2) = 0$ (why?), it follows that

$$\mathrm{Tr}(x) = x_0. \tag{11.11}$$

Furthermore $x^3 = x \cdot x^2 = (x_0 + x_1\alpha + x_2\alpha^2)(x_0 + x_2\alpha + (x_1 + x_2)\alpha^2) = (x_0 + x_1 + x_2 + x_1x_2) + (x_1 + x_0x_2 + x_0x_1)\alpha + (x_2 + x_0x_1)\alpha^2$. It follows then from (11.11) that

$$\mathrm{Tr}(x^3) = x_0 + x_1 + x_2 + x_1x_2. \tag{11.12}$$

From (11.12) we can analyze the sums (11.9) as follows. The function $\mathrm{Tr}(\beta x)$ is a linear function, i.e.,

$$\mathrm{Tr}(\beta x) = b_0x_0 + b_1x_1 + b_2x_2, \tag{11.13}$$

for certain $b_0, b_1, b_2 \in GF(2)$. Hence, using (11.12) we have

$$\mathrm{Tr}(\beta x + x^3) = a_0x_0 + a_1x_1 + a_2x_2 + x_1x_2, \tag{11.14}$$

where $a_i = b_i + 1$, $i = 0, 1, 2$. Alternatively,

$$\text{Tr}(\beta x + x^3) = a_0 y_0 + y_1 y_2 + a_1 a_2, \tag{11.15}$$

where $y_0 = x_0$, $y_1 = x_1 + a_2$, $y_2 = x_2 + a_1$. Now the sum (11.9) is a function only of the element β of $GF(8)$. There is a one-to-one correspondence between these eight β's and the eight binary vectors (a_0, a_1, a_2):

$$\beta = (a_0 + 1) + (a_1 + 1)\alpha + (a_2 + 1)\alpha^2,$$

with the element $\beta = 0$ corresponding to the vector $(1, 1, 1)$. For a fixed value of (a_0, a_1, a_2) (which corresponds to a fixed β, and, ultimately, to a fixed value of τ) there is a one-to-one correspondence between the eight elements $x \in GF(8)$ and the eight binary vectors (y_0, y_1, y_2):

$$x = y_0 + (y_1 + a_2)\alpha + (y_2 + a_1)\alpha^2.$$

Therefore to evaluate the function $C(\tau)$, for $\tau = 0, 1, \ldots, 6$, we must discover, for each of the seven vectors $(a_0, a_1, a_2) \neq (1, 1, 1)$, how many times the function (11.15) of the three binary variables y_0, y_1, y_2 is 0 and how many times it is 1. If this function is equal to 0 N times, and equal to 1 M times, then we know that the function $\text{Tr}(\beta x + x^3)$ will also equal 0 N times and 1 M times. It follows that the sum in (11.9), i.e., $C(\tau)$, which extends over only the *nonzero* values of β, will be equal to $(N - 1) - M$. Rather than consider all seven possibilities for (a_0, a_1, a_2) separately, we will distinguish just two cases, $a_0 = 1$ and $a_0 = 0$.

- Case 1: $a_0 = 1$. Then for any choice of y_1, y_2, the function $\text{Tr}(\beta x + x^3) = y_0 + y_1 y_2 + a_1 a_2$ (see (11.15)) will be 0 for one choice of y_0 and 1 for the other choice of y_0. Hence, for $x \neq 0$, $\text{Tr}(\beta x + x^3)$ will be 0 3 times, 1 4 times, and so $C(\tau) = -1$. There are three possible vectors (a_0, a_1, a_2) corresponding to $\beta \neq 0$ in this case, viz., (100), (101), (110). Hence in Case 1, we get a value of $C(\tau) = -1$, three times.
- Case 2: $a_0 = 0$. Here the function (11.15) becomes $\text{Tr}(\beta x + x^3) = y_1 y_2 + a_1 a_2$. To see how often this is zero, we must distinguish two sub-cases.
- Case 2a: $a_1 a_2 = 0$. Then $\text{Tr}(\beta x + x^3) = y_1 y_2$, which equals 0 six times and equals 1 twice. Hence $C(\tau) = (6 - 1) - 2 = 3$. There are 3 a-vectors in this subcase, viz., (000), (001), (010). Hence in Case 2a we have $C(\tau) = 3$, three times.

• Case 2b: $a_1 a_2 = 1$. Here $\mathrm{Tr}(\beta x + x^5) = y_1 y_2 + 1$, which equals 0 twice and 1 six times. Hence $C(\tau) = (2-1) - 6 = -5$. There is only one a-vector in this sub-case, viz., (011). Hence in Case 2b we get $C(\tau) = -5$, once.

In summary: we have accounted for all 7 values of τ, and obtained the following histogram.

value of $C(\tau)$	number of times attained	(a_0, a_1, a_2)
-5	1	(011)
-1	3	$(100), (101), (110)$
$+3$	3	$(000), (001), (010)$

This histogram is in agreement with the table in Example 11.1, as it should be. This may seem like a lot of work just to get the values of $C(\tau)$ in this one case, but this approach can be greatly generalized, as we will see. ∎

We now begin a serious study of sums of the form (11.9) in general. If the equation $\mathrm{Tr}(\beta x + x^d) = 0$ has N solutions in $GF(2^m)$, the summand in (11.9) will be $+1$ $N-1$ times and -1 $2^m - N$ times. It follows then that $C(\tau) = (N-1) - (2^m - N) = 2N - 2^m - 1$. So in order to evaluate the sum it will be sufficient to find the number of solutions to

$$F(x) = 0,$$

where $F(x) = \mathrm{Tr}(\beta x + x^d) = \mathrm{Tr}(\beta x) + \mathrm{Tr}(x^d)$. The function $\mathrm{Tr}(\beta x)$ is linear, and it so can be represented as

$$\mathrm{Tr}(\beta x) = \mathbf{x} \cdot \mathbf{b}$$

where $\mathbf{x}$ is the vector representation of x and $\mathbf{b} = (b_0, b_1, \ldots, b_{m-1})$ is a particular nonzero vector. The other part of $F(x)$, viz., $\mathrm{Tr}(x^d)$ is in general more complicated, but it is not perhaps as complicated as it appears. Let $d = d_0 + 2d_1 + 4d_2 + \cdots + 2^{m-1} d_{m-1}$ be the binary expansion of d (each d_k is 0 or 1). Then

$$x^d = \prod_{k=0}^{m-1} (x^{2^k})^{d_k}. \tag{11.16}$$

For any fixed integer k, the mapping $x \to x^{2^k}$ of $GF(2^m)$ onto itself is linear (Lemma 5.12), and so in terms of components of the vector $\mathbf{x} = (x_0, \ldots, x_{m-1})$, it is of the form

$$\mathbf{x} \to \mathbf{x}Q_k,$$

where Q_k is a nonsingular $m \times m$ matrix. Thus each component of x^{2^k} is a linear function of the components of $\mathbf{x}$. It follows then from (11.16) that the function $x \to x^d$ is the product of $w(d)$ linear functions of x, where $w(d)$ denotes the number of ones in the binary expansion of d (this is often called the *binary weight* of d). Since the trace operator is just a dot product, this means that each of the m binary components of $\mathrm{Tr}(x^d)$ is a Boolean function of degree at most $w(d)$ of the m Boolean variables $x_0, x_1, \ldots, x_m$.

The theory of polynomials of degree 1 in $x_0, x_1, \ldots, x_{m-1}$ is almost trivial, and the theory of polynomials of degree ≥ 3 is almost impossible; but *quadratic* polynomials are neither trivial nor impossible, and we will be able to make a lot of progress when d has the form $d = 1 + 2^e$, i.e., when $\mathrm{Tr}(x^d)$ is a quadratic function.

We must now pause for a somewhat general discussion of *quadratic forms.*

Let F be an arbitrary field, and let $x_1, x_2, \ldots, x_m$ be indeterminates over F. A *quadratic form* over F is a function of m variables $x_1, x_2, \ldots, x_m$ which can be expressed in the form

$$Q(x_1, x_2, \ldots, x_m) = \sum_{\substack{i,j=1 \\ i \leq j}}^{m} a_{ij} x_i x_j. \tag{11.17}$$

Such a form is called *nonsingular* if it cannot be transformed by a nonsingular change of variables into a form in fewer than m variables. For example, the form $x_1x_2 + x_1x_3$ in three variables is singular because the transformation $x_1 \leftarrow x_1$, $x_2 \leftarrow x_2 - x_3$, $x_3 \leftarrow x_3$ changes it to x_1x_2, a form involving only 2 variables. Informally, a quadratic form is nonsingular if it really is a form in m variables and not a disguised form of fewer variables.

A quadratic form $Q(x_1, x_2, \ldots, x_m)$ is said to *represent zero* if there exists $(\xi_1, \xi_2, \ldots, \xi_m) \neq (0, 0, \ldots, 0)$ such that $Q(\xi_1, \ldots, \xi_m) = 0$. The following theorem will be the key to our understanding of quadratic forms.

Theorem 11.2. *If Q is a nonsingular quadratic form as in (11.17), and if Q represents zero, then under a suitable nonsingular linear transformation Q can be put into the shape*

$$Q = x_1 x_2 + Q'(x_3, x_4, \ldots, x_m),$$

where Q' is a nonsingular quadratic form in $x_3, x_4, \ldots, x_m$.

Proof: Suppose $Q(\xi_1, \ldots, \xi_m) = 0$, not all of the ξ_i being zero. Consider any nonsingular linear transformation of the form

$$x_i \leftarrow \xi_i x_1 + \cdots, \qquad i = 1, 2, \ldots, m.$$

After this linear transformation has been performed, the coefficient of x_1^2 in the new form is clearly

$$\sum_{i,j} a_{ij} \xi_i \xi_j = 0.$$

Thus Q can be transformed into the shape

$$\begin{aligned} \text{(11.18)} \qquad & a'_{12} x_1 x_2 + a'_{13} x_1 x_3 + a'_{14} x_1 x_4 + \cdots \\ & + a'_{22} x_2^2 + a'_{23} x_2 x_3 + a'_{24} x_2 x_4 + \cdots \\ & + a'_{33} x_3^2 + a'_{34} x_3 x_4 + \cdots \end{aligned}$$

Not all of the a'_{1j}'s can be zero, or Q would be a function of only the $m-1$ variables $x_2, \ldots, x_m$. Thus (after a suitable permutation of coordinates, which is a special kind of linear transformation), we can assume $a_{12} \neq 0$. Then the linear transformation

$$\begin{aligned} x_2 &\leftarrow \frac{1}{a'_{12}}(x_2 - a'_{13} x_3 - a'_{14} x_4 - \cdots) \\ x_i &\leftarrow x_i \qquad (i \neq 2), \end{aligned}$$

will change the quadratic form (11.18) into the form

$$
\begin{aligned}
& x_1x_2 + a''_{22}x_2x_2 + a''_{23}x_2x_3 + \cdots \\
& \qquad\qquad\qquad + a'_{33}x_3^2 + \cdots
\end{aligned} \tag{11.19}
$$

Finally, the linear transformation

$$
\begin{aligned}
x_1 &\leftarrow x_1 - a''_{22}x_2 - a''_{23}x_3 - \cdots \\
x_i &\leftarrow x_i \qquad (i \neq 1)
\end{aligned}
$$

puts (11.19) into the form

$$
\begin{aligned}
& x_1x_2 + a'_{33}x_3^2 + \cdots \\
&= x_1x_2 + Q'(x_3, \ldots, x_m).
\end{aligned}
$$

Of course Q' must be nonsingular because if it were a function of fewer than $m-2$ variables the equation $Q = x_1x_2 + Q'$ could imply that Q itself was a function of fewer than m variables. ∎

Corollary 11.3. *With a suitable linear transformation, any nonsingular quadratic form of m variables can be put in the shape*

$$Q = x_1x_2 + x_3x_4 + \cdots + x_{2s-1}x_{2s} + Q'(x_{2s+1}, \ldots, x_m),$$

where Q' is a nonsingular form in $m-2s$ variables that does not represent zero.

In view of Corollary 11.3, our study of quadratic forms will not be complete until we can classify those quadratic forms which *do not* represent zero. In a general field this is not possible. But over finite fields it is a practical undertaking, because we shall now show that over a finite field *any quadratic polynomial in three or more variables must represent zero.* This fact is a special case of the famous theorem of Chevalley-Warning (Theorem 11.4 below). Before stating this theorem, however, we give a formal definition of a polynomial in many variables over a field F.

A *monomial* in the indeterminates $x_1, x_2, \ldots, x_m$ is an expression of the form $\lambda x_1^{e1} x_2^{e2} \cdots x_m^{em}$, where $\lambda \in F$ and the e_i are nonnegative integers. If some $e_i = 0$, the corresponding x_i is usually omitted; for example we write $\lambda x_1^2 x_4^7$ instead of $\lambda x_1^2 x_2^0 x_3^0 x_4^7$. The *degree* of such a monomial is defined to be $\sum e_i$. For example, the degree of $\lambda x_1^2 x_4^7$ is 9.

A *polynomial* in the indeterminates $x_1, x_2, \ldots, x_m$ is a sum of monomials. Its degree is defined to be the maximum degree of any of its monomials. Thus $1 + x_1^2 x_4^7 - x_1 x_2 x_3^9$ is a polynomial of degree 11.

Now suppose that F is a finite field with q elements. Given m fixed elements $\mathbf{a} = (a_1, a_2, \ldots, a_m)$ of F, we define the polynomial $P_{\mathbf{a}}$ as follows.

$$P_{\mathbf{a}}(x_1, \ldots, x_m) = \prod_{i=1}^{m} (1 - (x_i - a_i)^{q-1}) \tag{11.20}$$

It is easy to verify that this polynomial has the following useful *interpolation* property.

$$P_{\mathbf{a}}(x_1, \ldots, x_m) = \begin{cases} 1, & \text{if } (x_1, \ldots, x_m) = (a_1, \ldots, a_m) \\ 0, & \text{otherwise.} \end{cases} \tag{11.21}$$

Furthermore, the degree of $P_{\mathbf{a}}$ is $m(q-1)$, and its degree in each variable x_i separately is $q - 1$. If $f(x_1, \ldots, x_m)$ is any function mapping $F^m \to F$, we can represent it as a polynomial of degree $\leq m(q-1)$, using the interpolation polynomials (11.20):

$$f(\mathbf{x}) = \sum_{\mathbf{a} \in F^m} f(\mathbf{a}) P_{\mathbf{a}}(\mathbf{x}). \tag{11.22}$$

Furthermore, this representation of $f(x_1, \ldots, x_m)$ as a sum of monomials of the form $x_1^{e1} \cdots x_m^{em}$ with $0 \leq e_i \leq q-1$ is unique, since there are only q^{q^m} polynomials which can be formed from such polynomials, which is also the total number of functions mapping $F^m \to F$.

We can now state the theorem of Chevalley-Warning.

Theorem 11.4. *If $F = GF(q)$, where q is a power of the prime p, and if $f(x_1, \ldots, x_m)$ is a polynomial of degree $d < m$, then the number $N(f)$ of* *solutions to*

$$f(x_1, x_2, \ldots, x_m) = 0,$$

with $x_1, x_2, \ldots, x_m \in F$, *is divisible by* p.

Proof: For each m-tuple $\mathbf{x} \in F^m$, we have

$$1 - f(\mathbf{x})^{q-1} = \begin{cases} 1, & \text{if } f(\mathbf{x}) = 0 \\ 0, & \text{otherwise.} \end{cases}$$

Thus summing $1 - f(\mathbf{x})^{q-1}$ over all $\mathbf{x} \in F^m$, we get

$$N(f) \bmod p = \sum_{\mathbf{x}} (1 - f(\mathbf{x})^{q-1}) = -\sum_{\mathbf{x}} f(\mathbf{x})^{q-1},$$

Thus in order to prove the theorem it will be sufficient to prove that for any polynomial f with degree $< m$, we have

$$\sum_{\mathbf{x} \in F^m} f(\mathbf{x})^{q-1} = 0.$$

Now $f(\mathbf{x})^{q-1}$ is a polynomial of degree $d(q-1)$ and so is a linear combination of monomials of degree at most $d(q-1)$. If $m(\mathbf{x}) = x_1^{e_1} \cdots x_m^{e_m}$ is one such monomial, then

$$\sum_{\mathbf{x}} m(\mathbf{x}) = \prod_{i=1}^{m} \sum_{x \in F} x^{e_i}. \tag{11.23}$$

If any $e_i = 0$, the corresponding sum in (11.23) will be zero. On the other hand, since $e_1 + \cdots + e_m \leq d(q-1) < m(q-1)$, unless all the e_i's are zero, one of the e_i's will be in the range $1 \leq e_i < q-1$. If α is a primitive root in F, the sum $\sum_{x \in F} x^{e_i}$ in (11.23) is then

$$\sum_{j=0}^{q-2} \alpha^{je_i} = \frac{\alpha^{e_i(q-1)} - 1}{\alpha^{e_i} - 1} = 0.$$

Thus in every case the sum (11.23) is zero. ∎

Corollary 11.5. *Over any finite field, a quadratic form in $m \geq 3$ variables represents zero.*

Proof: Let $Q = \sum a_{ij}x_i x_j$. Then $Q(0, 0, \ldots, 0) = 0$, and so $N(f)$ cannot be 0. On the other hand, Theorem 11.4 tells us that $N(f)$ is divisible by p, the characteristic of F, and so is at least p. Hence there must be at least $p-1$ additional zeros, i.e., at least $p-1$ nonzero vectors for which Q is zero. ∎

We now specialize to the binary field $GF(2)$. The only quadratic form in one variable over $GF(2)$ is

$$Q_1(x) = x^2 = x.$$

Plainly Q_1 does not represent 0. In two variables there are four quadratic forms, viz., $x^2 + xy$, $x^2 + y^2$, $xy + y^2$, $x^2 + xy + y^2$. One easily checks that the first three represent zero, and so the only quadratic form over $GF(2)$ in two variables that does not represent zero is

$$\begin{aligned} Q_2(x, y) &= x^2 + xy + y^2 \\ &= x + xy + y \end{aligned}$$

These remarks, combined with Theorems 11.2 and 11.3, prove the following theorem.

Theorem 11.6. *Every nonsingular quadratic form in m variables over $GF(2)$ is equivalent, under a linear transformation of the variables, to exactly one of the following:*

(m odd):

$$x_1x_2 + x_3x_4 + \cdots + x_{m-2}x_{m-1} + x_m$$

(m even):

$$x_1x_2 + x_3x_4 + \cdots + x_{m-1}x_m,$$

or

$$x_1x_2 + x_3x_4 + \cdots + x_{m-1}x_m + x_{m-1} + x_m.$$

Corollary 11.7. *Every quadratic form in m variables (nonsingular or not) over $GF(2)$ is equivalent to exactly one of the following:*

$$x_1x_2 + x_3x_4 + \cdots + x_{2s-1}x_{2s} + x_{2s+1} \quad (\text{rank } 2s+1)$$
$$x_1x_2 + x_3x_4 + \cdots + x_{2s-1}x_{2s} \quad (\text{rank } 2s)$$
$$x_1x_2 + x_3x_4 + \cdots + x_{2s-1}x_{2s} + x_{2s-1} + x_{2s} \quad (\text{rank } 2s)$$

where $s = \lfloor r/2 \rfloor$, where r is the rank of the quadratic form in question.

Proof: Let $Q(x_1, \ldots, x_m)$ be a given quadratic form. Denote by r the minimum number of variables that Q can be expressed in terms of by a nonsingular linear transformation of variables. The number r is called the *rank* of Q. Thus $Q(x_1, \ldots, x_m) = Q'(x'_1, x'_2, \ldots, x'_r)$. Clearly Q' must be a nonsingular quadratic form in the new variables x'_i. Now apply Theorem 11.6. ■

Corollary 11.7 allows us to establish the following useful fact about the rank of a quadratic form over $GF(2)$.

Corollary 11.8. *If $Q(x_1, x_2, \ldots, x_m)$ is a quadratic form over $GF(2)$ of rank r, then the number of m-tuples $(b_1, b_2, \ldots, b_m)$ such that*

$$Q(a_1 + b_1, a_2 + b_2, \ldots, a_m + b_m) = Q(a_1, a_2, \ldots, a_m) \tag{11.24}$$

for all 2^m m-tuples $(a_1, a_2, \ldots, a_m)$ is equal to 2^{m-r}.

Proof: We use Corollary 11.7. If r is odd, then Q can be transformed into the form $x_1x_2 + x_3x_4 + \cdots + x_{2s-1}x_{2s} + x_{2s+1}$, and the condition (11.24) becomes

$$\begin{aligned}(a_1b_2 + a_2b_1 + b_1b_2) + \cdots + (a_{2s-1}b_{2s} + a_{2s}b_{2s-1} + b_{2s-1}b_{2s})\\ + b_{2s+1} = 0\end{aligned} \tag{11.25}$$

for all $(a_1, a_2, \ldots, a_m)$. In particular, if $(a_1, a_2, \ldots, a_m) = (0, 0, \ldots, 0)$, (11.25) implies that

$$b_1b_2 + \cdots + b_{2s-1}b_{2s} + b_{2s+1} = 0. \tag{11.26}$$

Combining (11.26) with (11.25), we conclude that

$$(a_1 b_2 + a_2 b_1) + \cdots + (a_{2s-1} b_{2s} + a_{2s} b_{2s-1}) = 0 \tag{11.27}$$

for all $(a_1, a_2, \ldots, a_m)$. This can only be true if $b_1 = b_2 = \cdots = b_{2s} = 0$, and this, combined with (11.26), implies that also $b_{2s+1} = 0$. In summary, we have shown that (11.25) holds for all $a_1, a_2, \ldots, a_m$ if and only if $b_1 = b_2 = \cdots = b_{2s+1} = 0$. There are exactly 2^{m-2s-1} such m-tuples, and since $2s+1$ is the rank of Q, this completes the proof for odd r. The case of even r is handled similarly, and will be left as a problem. ∎

Let us now return to our main goal, the calculation of the number of solutions in $GF(2^m)$ of the equation

$$F(x) = \mathrm{Tr}(x^{1+2^e}) + \mathrm{Tr}(\beta x) = 0. \tag{11.28}$$

By Corollary 11.7 we see that $F(x) = 0$ can be transformed into an equation of the form

$$x_1 x_2 + \cdots + x_{2s-1} x_{2s} + a_1 x_1 + a_2 x_2 + \cdots + a_m x_m = 0, \tag{11.29}$$

where $x_1, x_2, \ldots, x_m$ are allowed to assume only the values 0 and 1, and $s = \lfloor r/2 \rfloor$, where r is the rank of the quadratic form represented by $\mathrm{Tr}(x^{1+2^e})$. For a fixed value of s, there are 2^m equations of the form (11.29), one for each choice of the m values $(a_1, a_2, \ldots, a_m)$. The following theorem will tell us exactly how many solutions each of these equations has.

Theorem 11.9. *The number of solutions to (11.29) is given in the following table:*

No. of solutions	No. of equations
2^{m-1}	$2^m - 2^{2s}$
$2^{m-1} + 2^{m-s-1}$	$2^{2s-1} + 2^{s-1}$
$2^{m-1} - 2^{m-s-1}$	$2^{2s-1} - 2^{s-1}$

Proof: We distinguish two cases (cf. Example 11.2).

• Case 1: $a_i = 1$ for some $i > 2s$. There are $2^m - 2^{2s}$ such choices for $(a_1, \ldots, a_m)$. In this case (11.29) can be written in the form

$$f(x_1, x_2, \ldots, x_m) + x_i = 0, \tag{11.30}$$

where $f(x_1, \ldots, x_m)$ does not involve x_i. Then for any choice of the $m-1$ x's other than x_i, there will be exactly one choice for x_i such that (11.30) holds. It follows that the number of solutions to (11.29) is 2^{m-1} in this case. This accounts for the first entry in the table.

• Case 2: $a_{2s+1} = a_{2s+2} = \cdots = a_m = 0$. There are a total of 2^{2s} such choices for $(a_1, \ldots, a_m)$. In this case the transformation $y_1 = x_1 + a_2$, $y_2 = x_2 + a_1$, $y_3 = x_3 + a_4$, $y_4 = x_4 + a_3$, etc., changes (11.29) into

$$y_1 y_2 + \cdots + y_{2s-1} y_{2s} = a, \tag{11.31}$$

where $a = a_1 a_2 + a_3 a_4 + \cdots + a_{2s-1} a_{2s}$. Now a can assume only the values 0 and 1, and so it is sufficient to consider only the simpler equation

$$y_1 y_2 + \cdots + y_{2s-1} y_{2s} = 0. \tag{11.32}$$

Denote by N_s the number of solutions $(y_1, y_2, \ldots, y_{2s})$ to (11.32). By actually counting, perhaps, we find that

$$N_1 = 3, N_2 = 10, N_3 = 36, \ldots \tag{11.33}$$

To find a general formula for N_s, we proceed by induction. If N_s is the number of solutions to (11.32), then N_{s+1} is the number of solutions to

$$y_1 y_2 + \cdots + y_{2s-1} y_{2s} = y_{2s+1} y_{2s+2}. \tag{11.34}$$

If the left side of (11.34) is equal to zero, (N_s ways), the right side must also be zero (3 ways). But if the left side is one ($2^{2s} - N_s$ ways), the right must be one (1 way). Hence

$$N_{s+1} = 3(N_s) + (2^{2s} - N_s),$$

or equivalently,

$$N_{s+1} - 2N_s = 2^{2s}. \tag{11.35}$$

Using (11.35) and the initial conditions (11.33), one can verify that in fact

$$N_s = 2^{2s-1} + 2^{s-1}. \tag{11.36}$$

It follows from (11.35) that the number of solutions $(y_1, y_2, \ldots, y_m)$ to (11.29) is given by

(11.37*a*) $\quad 2^{m-1} + 2^{m-s-1} \quad$ if $a = 0$

(11.37*b*) $\quad 2^{m-1} - 2^{m-s-1} \quad$ if $a = 1$,

where $a = a_1a_2 + \cdots + a_{2s-1}a_{2s}$. But there are exactly N_s choices for $(a_1, \ldots, a_{2s})$ such that $a = 0$, and $2^{2s} - N_s$ such that $a = 1$. This accounts for the second and third entries in the statement of the theorem, and completes the proof. ∎

Theorem 11.9 will allow us to calculate the complete histogram of values assumed by the crosscorrelation function $C(\tau)$ in the case $d = 1 + 2^e$, provided only that we can identify the parameter s that appeared in the transformation from (11.28) to (11.29). But we already noted that $s = \lfloor r/2 \rfloor$, where r is the rank of the quadratic form

$$Q_e(x) = \mathrm{Tr}(x^{1+2^e}). \tag{11.38}$$

The next theorem will identify the rank of $Q_e(x)$.

Lemma 11.10. *The rank of the quadratic form $Q_e(x)$ in (11.38) is:*

$$\mathrm{rank}(Q_e) = \begin{cases} m - \gcd(m, 2e), & \text{if } \gcd(m, 2e) = 2\gcd(m, e) \\ m - \gcd(m, 2e) + 1, & \text{if } \gcd(m, 2e) = \gcd(m, e). \end{cases}$$

Proof: We will apply Corollary 11.8. In order to do so, we need to calculate the size of the set Y_e, defined by

(11.39)
$$Y_e = \{y \in GF(2^m) : \mathrm{Tr}((x+y)^{2^e+1}) = \mathrm{Tr}(x^{2^e+1}), \quad \text{for all } x \in GF(2^m)\}.$$

Consider the following computation:

$$\begin{aligned}\mathrm{Tr}((x+y)^{2^e+1}) &= \mathrm{Tr}((x+y)^{2^e}(x+y)) \\ &= \mathrm{Tr}((x^{2^e}+y^{2^e})(x+y)) \\ &= \mathrm{Tr}(x^{2^e+1}+x^{2^e}y+xy^{2^e}+y^{2^e+1}) \\ &= \mathrm{Tr}(x^{2^e+1})+\mathrm{Tr}(x^{2^e}y)+\mathrm{Tr}(xy^{2^e})+\mathrm{Tr}(y^{2^e+1}).\end{aligned}$$

Now since $\mathrm{Tr}(\alpha) = \mathrm{Tr}(\alpha^{2^e})$, the term $\mathrm{Tr}(xy^{2^e})$ equals $\mathrm{Tr}(x^{2^e}y^{2^{2e}})$, and so

$$\mathrm{Tr}((x+y)^{2^e+1}) = \mathrm{Tr}(x^{2^e+1}+x^{2^e}(y+y^{2^{2e}})+y^{2^e+1}).$$

Hence the equation

$$\mathrm{Tr}((x+y)^{2^e+1}) = \mathrm{Tr}(x^{2^e+1})$$

is equivalent to

$$\mathrm{Tr}(x^{2^e}(y+y^{2^{2e}})) = \mathrm{Tr}(y^{2^e+1}). \tag{11.40}$$

and the set Y_e of (11.39) is the set of $y \in GF(2^m)$ such that (11.40) holds for all $x \in GF(2^m)$.

As x runs through all elements of $GF(2^m)$, so does x^{2^e}. Thus if $y + y^{2^e} \neq 0$, as x runs through $GF(2^m)$, the left side of (11.40) is 0 2^{m-1} times and 1 2^{m-1} times. But the right side of (11.40) is does not depend on x, and this is a contradiction. Hence if (11.40) is satisfied for all $x \in GF(2^m)$,

$$y = y^{2^{2e}}. \tag{11.41}$$

Now if (11.41) holds, the left side of (11.40) is identically zero. Hence in order for (11.40) for all x, we must have, in addition to (11.41),

$$\mathrm{Tr}(y^{2^e+1}) = 0. \tag{11.42}$$

Therefore the set Y_e of (11.39) is exactly the set of y's such that (11.41) and (11.42) hold.

We know that (11.41) is by Lemma 5.10 equivalent to $y \in GF(2^{2e})$; since also $y \in GF(2^m)$, we therefore have (see Eq. (6.22))

$$y \in GF(2^{\gcd(2e,m)}). \tag{11.43}$$

Since $y \in GF(2^{2e})$, if $y \neq 0$, we have $y^{2^{2e}-1} = 1$, i.e., $y^{(2^e+1)(2^e-1)} = 1$. This in turn implies that $y^{2^e+1} \in GF(2^e)$, but since also $y \in GF(2^m)$, we have

$$y^{2^e+1} \in GF(2^{\gcd(e,m)}). \tag{11.44}$$

We must now distinguish two cases, according to whether the fields appearing on the right sides of (11.43) and (11.44) are the same or different. To simplify the notation, we introduce the symbols g and h:

$$g = \gcd(2e, m) \tag{11.45}$$

$$h = \gcd(e, m). \tag{11.46}$$

The possibilities are that either $g = 2h$ or $g = h$.

- Case 1: $g = 2h$. In this case it follows that $\mathrm{Tr}(y^{2^e+1}) = 0$ for all $y \in GF(2^g)$. This is because according to Theorem 8.2, for any $z \in GF(2^m)$,

$$\mathrm{Tr}(z) = \mathrm{Tr}_1^h(\mathrm{Tr}_h^m(z)), \tag{11.47}$$

where in (11.45) Tr_1^h denotes the trace from $GF(2^h)$ to $GF(2)$, and Tr_h^m denotes the trace from $GF(2^m)$ to $GF(2^h)$. Now for any $z \in GF(2^h)$, $z^{2^h} = z$ and so

$$\begin{aligned}\mathrm{Tr}_h^m(z) &= z + z + \cdots + z \\ &= \tfrac{m}{h} \cdot z.\end{aligned}$$

But in this case (Case 1), we know that m/h is even, since

$$\tfrac{m}{h} = \tfrac{m}{g} \cdot \tfrac{g}{h} = 2\tfrac{m}{g}.$$

Therefore every element of $GF(2^h)$, and in particular y^{2^e+1} (because of (11.44)), has trace 0, and so the set Y_e is the subfield $GF(2^g)$. It follows then from Corollary 11.8 that $\text{rank}(Q_e) = m - g = m - \gcd(2e, m)$ in Case 1.

• Case 2: $g = h$. In this case the field $GF(2^g)$ and $GF(2^h)$ are the same, and $\text{Tr}(y^{2^e+1})$ will not be identically zero. But in this case we have

$$\gcd(2^e + 1, 2^h - 1) \mid \gcd(2^e + 1, 2^e - 1),$$

and $\gcd(2^e + 1, 2^e - 1) = 1$, since the two numbers differ by 2 and both are odd. It follows that as y runs through the elements of $GF(2^h)$, y^{2^e+1} does also, and so exactly *half* the elements of $GF(2^h)$ will satisfy (11.42). In this case we conclude that the number of elements $y \in Y_e$ is 2^{h-1}, and hence $\text{rank}(Q_e) = m - h + 1 = m - \gcd(e, m) + 1 = m - \gcd(2e, m) + 1$. ∎

Let us recapitulate. We are trying to determine the values of the sums described in (11.9) when $d = 2^e + 1$. When $\gcd(2e, m) = \gcd(e, m)$ these sums will represent the crosscorrelation function $C(\tau)$ between a given m-sequence and its dth decimation. The sums in (11.9) are parameterized by the element $\beta \in GF(2^m)$, where $\beta = \alpha^{-\tau}$, and α is a fixed primitive root in $GF(2^m)$. The sum in (11.9) is equal to $2N - 2^m - 1$, where N is the number of solutions to the equation (11.28). In Theorem 11.9 we found that there were only three possibilities for N, and calculated the exact number of values of β that gave rise to each of these three possibilities. However, the results of Theorem 11.9 were stated in terms of the parameter s, which is equal to $\lfloor r/2 \rfloor$, where r is the rank of the quadratic form Q_e defined in (11.38). Finally, in Lemma 11.10, we determined the rank of Q_e as $m - g$ or $m - g - 1$, where $g = \gcd(2e, m)$. If we note that $m - g$ is always even, it follows that the parameter s in Theorem 11.9 is equal to $(m - g)/2$. Putting all this together, we obtain the following table, which gives the three possible values for $C(\tau)$, and *almost* gives the number of times each possible value is assumed by $C(\tau)$.

$C(\tau)$	No. of times assumed
-1	$2^m - 2^{m-g}$
$-1 + 2^{(m+g)/2}$	$2^{m-g-1} + 2^{(m-g)/2-1}$
$-1 - 2^{(m+g)/2}$	$2^{m-g-1} - 2^{(m-g)/2-1}$

We say almost, because included in this table is the contribution of the sum (11.9) with $\beta = 0$, which does not correspond to any value of τ. To be absolutely complete, we must determine the value of this sum and remove it from the table. The next theorem will allow us to do this.

Theorem 11.11. *The number of solutions in $GF(2^m)$ of the equation*

$$\mathrm{Tr}(x^{2^e+1}) = 0 \tag{11.48}$$

assumes one of three values, as described below.

- *Case 1:* 2^{m-1} *if $g = h$,*
- *Case 2:* $2^{m-1} + 2^{(m+g)/2-1}$ *if $g = 2h$ and m/g is odd,*
- *Case 3:* $2^{m-1} - 2^{(m+g)/2-1}$ *if $g = 2h$ and m/g is even.*

Proof: We know that the quadratic form $Q_e(x) = \mathrm{Tr}(x^{2^e+1})$ can be transformed into exactly one of the three types of quadratic forms described in Corollary 11.7, viz.

$$x_1x_2 + \cdots + x_{2s-1}x_{2s} + x_{2s+1}$$

$$x_1x_2 + \cdots + x_{2s-1}x_{2s}$$

$$x_1x_2 + \cdots + x_{2s-1}x_{2s} + x_{2s-1} + x_{2s}.$$

In the first of these three cases, we know from the proof of Theorem 11.9 that the number of solutions to $Q_e(x) = 0$ is 2^{m-1}; furthermore since the rank $2s + 1$ is odd, Lemma 11.10 says that this happens only when $g = h$. This justifies Case 1 in the statement of the present theorem.

If however the rank of Q_e is even, i.e., if $g = 2h$, then all we can conclude from the proof of Theorem 11.9 is that the number of solutions to $Q_e(x) = 0$ is either $2^{m-1} + 2^{(m+g)/2-1}$ or $2^{m-1} - 2^{(m+g)/2-1}$. To distinguish these two possibilities we need a lemma.

Lemma 11.12. *The number of nonzero solutions to (11.48) must be a multiple of $d = \gcd(2^e + 1, 2^m - 1)$, which, by Lemma 11.1, is $2^h + 1$.*

Proof: The mapping $x \to x^{2^e+1}$ of the nonzero elements of $GF(2^m)$ is a homomorphism and its kernel is the set $K_e = \{x : x^{2^e+1} = 1\} = \{y : y^d = 1\}$. This kernel contains exactly d elements, and so any element which is covered at all in the mapping is covered a multiple of d times. In particular, the number of nonzero solutions to (11.48) equals d times the number of elements in the range of the mapping with zero trace. ∎

With the help of Lemma 11.12 we can complete the proof of Theorem 11.11. We now know two things about the number of nonzero solutions to (11.48), when $g = 2h$. First, we know that this number is

$$2^{m-1} \pm 2^{(m+g)/2-1} - 1,$$

but we are uncertain as to whether the sign is plus or minus. Second, we know by Lemma 11.12 that this number is a multiple of $2^h + 1$. Combining these facts, we obtain the congruence

$$2^m \pm 2^{(m+g)/2} \equiv 2 \pmod{2^h + 1}. \tag{11.49}$$

Now $m = 2h(m/g)$ and so (11.49) becomes

$$(2^{2h})^{(m/g)} \pm (2^h)^{(m/g)+1} \equiv 2 \pmod{2^h + 1}. \tag{11.50}$$

But since $2^h \equiv -1 \pmod{2^h + 1}$, (11.50) in turn becomes

$$1 \pm (-1)^{(m/g)+1} \equiv 2 \pmod{2^h + 1}. \tag{11.51}$$

Finally we can see (11.51) can be true with a "+" sign only if m/g is odd, and with a "–" sign only if m/g is even. This takes care of Cases 2 and 3 of the theorem, and completes the proof. ∎

We are now prepared to state the main theorem of this chapter.

Theorem 11.13. *The histogram for the function $C(\tau)$, defined in (11.9), for $\tau = 0, 1, \ldots, 2^m - 2$, when $d = 2^e + 1$, is as given in the following table.*

$C(\tau)$	*No. of times assumed*	*(less one if)*	
-1	$2^m - 2^{m-g}$	$g = h$	
$-1 + 2^{(m+g)/2}$	$2^{m-g-1} + 2^{(m-g)/2-1}$	$g = 2h,$	m/g is odd
$-1 - 2^{(m+g)/2}$	$2^{m-g-1} - 2^{(m-g)/2-1}$	$g = 2h,$	m/g is even

∎

Corollary 11.14. *The histogram for the crosscorrelation function $C(\tau)$ between an m-sequence of length $2^m - 1$ and its $2^e + 1$st decimation (necessarily $\gcd(2^m - 1, 2^e + 1) = 1$) is given in the following table.*

$C(\tau)$	*No. of times assumed*
-1	$2^m - 2^{m-g} - 1$
$-1 + 2^{(m+g)/2}$	$2^{m-g-1} + 2^{(m-g)/2-1}$
$-1 - 2^{(m+g)/2}$	$2^{m-g-1} - 2^{(m-g)/2-1}$

∎

Example 11.3. We can illustrate Theorem 11.13 with a numerical example. We choose as our basic m-sequence one of length 63, so that $m = 6$. The following table summarizes the predictions of Theorem 11.13.

e	$2^e + 1$	g	h	$C(\tau)$	histogram
1	3	2	1	$-1, 15, -17$	$48, 9, 6$
2	5	2	2	$-1, 15, -17$	$47, 10, 6$
3	9	6	3	$-1, 65, 63$	$63, 0, 0$
4	17	2	2	$-1, 15, -17$	$47, 10, 6$
5	33	2	1	$-1, 15, -17$	$48, 9, 6$

First we note that $g = h$ in only two cases, viz. $e = 2$ and $e = 4$, so that only these two table entries correspond to $C(\tau)$ between two m-sequences. Furthermore it is easy to check that α^5 and α^{17} are conjugate in $GF(64)$, so that in fact the 5th and 17th decimations of the original m-sequence produce the same m-sequence, and indeed the entries for $e = 2$ and $m = 4$ are identical. Next, we note that for $e = 1$ we have $g = 2$ but $h = 1$, so that decimating a length 63 m-sequence by 3 will not produce an m-sequence. Indeed since 3 is a divisor of 63, we know that such a decimation will produce a sequence of period 21. Nevertheless, the given values do describe the crosscorrelation function between the original m-sequence and this period 21 sequence. Also, since α^3 and α^{33} are conjugate, the entries for $e = 1$ and $e = 5$ are identical. Finally, we note the peculiar case $e = 3$. In this case the decimation is by 9, which produces a sequence of period 7. In fact, this sequence is an m-sequence of length 7, but the entries in the table *do not* correspond to the values of the crosscorrelation function between this m-sequence and the original one. The reason is that the function $\mathrm{Tr}(x^9)$ in $GF(64)$ is identically zero, since x^9 will always lie in $GF(8)$, and the $GF(64)$-trace of every element in $GF(8)$ is zero. ∎

At the beginning of this long chapter we said that our motivation for studying the $C(\tau)$ problem was to find pairs of sequences for which the crosscorrelation function was uniformly low. And now we know, thanks to Theorem 11.12, that the crosscorrelation function between an m-sequence and its (2^e+1)th decimation can be as low as $2^{(m+1)/2}$, if $g = 1$. However, this fact will provide good signature sequences for only two users. If there are more than two users, Theorem 11.13 is apparently of no help. However, by a slight modification of our previous work, we can produce a large number of sequences such that the crosscorrelation function between each pair of them is small.

Thus again let $d = 2^e + 1$, and for each $x \in GF(2^m)$, define a sequence $s_t(x)$ of length $n = 2^m - 1$ as follows.

$$s_t(x) = \mathrm{Tr}(\alpha^t + x\alpha^{dt}) \qquad 0 \le t \le n-1. \tag{11.52}$$

There are 2^m such sequences, one for each $x \in GF(2^m)$. If by $x = \infty$ we mean the sequence

$$s_t(\infty) = \mathrm{Tr}(\alpha^{dt}) \tag{11.53}$$

we have altogether $2^m + 1$ such sequences. For any x, y, let us denote by $C_{xy}(\tau)$ the crosscorrelation function which compares $s_t(x)$ and $s_t(y)$:

$$C_{xy}(\tau) = \sum_{t=0}^{n-1} (-1)^{s_t(x)+s_{t+\tau}(y)}. \tag{11.54}$$

Now $s_t(x) + s_{t+\tau}(y) = \mathrm{Tr}(\alpha^t + \alpha^{t+\tau} + x\alpha^{dt} + y\alpha^{dt+d\tau})$, and so computing $C_{xy}(\tau)$ is essentially equivalent to finding the number of integers $0 \le t \le n-1$ such that

$$\mathrm{Tr}((1+\alpha^\tau)\alpha^t + (x + y\alpha^{d\tau})\alpha^{dt}) = 0.$$

If we denote $1+\alpha^\tau$ by A, $x + y\alpha^{d\tau}$ by B, and α^t by z, the problem is to find the number of nonzero $z \in GF(2^m)$ such that

$$\mathrm{Tr}(Az + Bz^d) = 0.$$

By an analysis exactly like what we have already done, we can show that the number of such solutions will be -1, $-1 \pm 2^{(m+g)/2}$, unless both A and B are zero, i.e.,

$$\begin{aligned} \alpha^\tau &= 1 \\ x + y\alpha^{d\tau} &= 0. \end{aligned}$$

But clearly this happens if and only if $\tau \equiv 0 \pmod{n}$, and $x = y$. Here is our conclusion:

Theorem 11.15. *If $d = 2^e+1$, then the 2^m+1 sequences defined by (11.52), have*

$$C_{xy}(\tau) = -1 + \epsilon 2^{(m+g)/2},$$

where ϵ is either 0, 1 or -1, unless $\tau \equiv 0 \pmod{n}$ and $x = y$, in which case $C_{xy}(\tau) = n$. ∎

The sequences described in Theorem 11.15 are called (generalized) *Gold sequences*, in honor of one of their earliest investigators. The theorem is strongest, i.e., gives the smallest uniform bound on $|C_{xy}(\tau)|$, when $g = 1$, i.e., when m is odd, and $\gcd(e, m) = 1$. These are the original Gold sequences, and are in widespread use in multiple-user communication systems.

Problems for Chapter 11.

1. Prove that the error estimate (11.2) is valid.

2. Calculate the odd crosscorrelation function $\overline{C}(\tau)$ between the two m-sequences 0010111 and 0011101 of length 7.

3. Calculate the crosscorrelation function between an m-sequence and *itself.*

4. Let $C(\tau)$ denote the crosscorrelation function for the two sequences $\mathbf{x}$ and $\mathbf{y}$, and let $C'(\tau)$ denote the crosscorrelation function for the shifted sequences $S^a\mathbf{x}$ and $S^b\mathbf{y}$. Find an expression for $C'(\tau)$ in terms of $C(\tau)$.

5. Let (s_t) and (r_t) be two (not necessarily distinct) m-sequences of length $n = 2^m - 1$. (Assume the components are ± 1.) Let $C(\tau)$ denote their crosscorrelation function.

a. Show that $\sum_{\tau=0}^{n-1} C(\tau) = 1$.

b. Find a formula for $\sum_{\tau=0}^{n-1} C(\tau)^2$ that depends only on m.

c. Check your results by computing $C(0), \ldots, C(14)$ for two distinct m-sequences of length 15.

6. This problem will discuss the crosscorrelation function for a pair of m-sequences of *different lengths.*

a. Let

$$(s_t) = (+1, -1, -1)$$

and

$$(r_t) = (+1, +1, -1, +1, -1, -1, -1).$$

(these are m-sequences of length 3 and 7, respectively.) Define $C(\tau) = \sum_{t=0}^{20} s_t r_{t+\tau}$, where both s_t and r_t are repeated periodically, and the subscripts are taken mod 21. Calculate $C(\tau)$ for all τ.

b. Generalize part (a) and compute the crosscorrelation function between an arbitrary pair of m-sequences whose periods are relatively prime.

7. Which values are possible for $C(\tau)$ for a length 31 m-sequence and its reverse?

8. In the course of the proof of Lemma 11.1, we made the following three assertions. Please prove them.

a. $2^e + 1$ and $2^e - 1$ are relatively prime.
b. $2^e + 1$ and $2^{\gcd(e,m)} - 1$ are relatively prime.
c. $e/\gcd(e,m)$ is odd, if $\gcd(2e,m) = 2\gcd(e,m)$.

9. Evaluate the function $C(\tau)$ in (11.9) when d is a power of two.

10. Prove that the following facts about the "interpolation polynomial" $P_{\mathbf{a}}$ defined in (11.20).

a. Its degree in the variable x_i is $q-1$.
b. Its total degree is $m(q-1)$.
c. It has the property described in (11.21).

11. Find, for each $m \geq 1$, a quadratic form over the field of real numbers that does not represent 0.

12. Verify that x^2+xy, x^2+y^2, and $xy+y^2$ all represent zero over $GF(2)$.

13. The object of this problem is for you to find an analog of Theorem 11.6 for the field $GF(3)$.

a. Find all quadratic forms $Q(x,y) = Ax^2 + Bxy + Cy^2$ in two variables over the field $GF(3)$ which do not represent zero.
b. State and prove a $GF(3)$-analog of Theorem 11.6.

14. The object of this problem is for you to classify those quadratic forms of the form $Q = Ax^2 + Bxy + Cy^2$ which do not represent 0 over the field $GF(2^m)$.

a. Show that such a form has $ABC \neq 0$.
b. Show that a linear transformation of the form $x \leftarrow \lambda x$, $y \leftarrow \mu y$ puts a Q with $ABC \neq 0$ into the form $\alpha x^2 + xy + \alpha y^2$.
c. Show that $\text{Tr}(\alpha) = 1$. [Hint: Review Theorem 8.4.]
d. Finally show that if λ is any fixed element in $GF(2^m)$ with trace 1, Q can be transformed to $\lambda x^2 + xy + \lambda y^2$.

15. Prove Corollary 11.8 for even r.

16. How many solutions (x_1, x_2, x_3, x_4) does the equation

$$x_1x_2 + x_3x_4 + x_1 + x_2 + x_3 + x_4 = 0$$

have?

17. Show that any equation of the form

$$f(x_1, x_2, \ldots, x_{m-1}) + \lambda x_m = 0,$$

where $\lambda \neq 0$, has exactly q^{m-1} solutions in the field $GF(q)$.

18. Let $P(x_1, x_2, \ldots, x_m)$ be a polynomial, not the zero polynomial, with coefficients in the finite field $k = GF(q)$, and suppose that P has degree $\leq q-1$ in each variable. Show that there exists $(\xi_1, \ldots, \xi_m)$ such that $P(\xi_1, \ldots, \xi_m) \neq 0$. [Hint: Use induction on m.]

19. Verify (11.36).

20. The object of this problem is for you to count the solutions to the equation

$$x_1x_2 + x_3x_4 + \cdots + x_{2m-1}x_{2m} = a,$$

where a is a fixed element in the finite field $GF(q)$.

a. Denoting by $N_m(a)$ the number of solutions, show that $N_m(a) = N_m(1)$ if $a \neq 0$.

b. By rewriting the equation as

$$x_1x_2 + \cdots + x_{2m-3}x_{2m-2} = a - x_{2m-1}x_{2m},$$

show that

$$N_m(0) = (2q-1)N_{m-1}(0) + (q-1)^2 N_{m-1}(1)$$
$$N_m(1) = (q-1)N_{m-1}(0) + (q^2-q+1)N_{m-1}(1).$$

c. Now solve for $N_m(0)$ and $N_m(1)$.

21. Show that $m - g = m - \gcd(2e, m)$ is always even.

22. Verify the entries in the table of values of $C(\tau)$ given just before the statement of Theorem 11.11.

23. Prove that every element in $GF(8)$ has trace zero in $GF(64)$.

24. Suppose m is even and $e = m/2$. What does Theorem 11.13 predict? Can you explain this rather peculiar prediction?

25. Supply the details of the proof of Theorem 11.15. Be alert for the special case $x = \infty$.

26. Find the least positive constant A such that the bound

$$\left| \sum_{x \in GF(q)} (-1)^{\mathrm{Tr}(x^{65} + \beta x)} \right| \leq A\sqrt{q}$$

is valid for all $\beta \in GF(q)$ and all q's of the form 2^m. Can you strengthen the bound for some values of q?

27. Take $m = 4$ and $e = 3$ in Theorem 11.15, and explicitly construct the Gold sequences.

Bibliography

There are very few books that are devoted in whole, or even in large part, to the theory of finite fields. (One of the reasons I decided to write this book is the scarcity of potential competition!) Still, *very few* is not the same as *none at all*, and in this brief bibliography I will list the five books that I have found to be useful references for engineering and computer science students wishing to learn about finite fields.

A. A. Albert, *Fundamental Concepts of Higher Algebra.* Chicago: University of Chicago Press, 1956. This little book contains one chapter (Chapter V, 32 pages long) devoted entirely to finite fields. It is especially good as a reference about polynomials over finite fields.

E. R. Berlekamp, *Algebraic Coding Theory.* Laguna Hills, Calif.: Aegean Park Press, 1984. (Reprint, with revisions, of the 1968 McGraw-Hill original.) The classic book on algebraic coding theory, including the Galois Gospel according to Berlekamp. It contains excellent discussions of digital circuitry to implement finite field arithmetic, Euclid's algorithm, Berlekamp's factorization algorithm, and much more.

L. E. Dickson, *Linear Groups, with an Exposition of the Galois Field Theory.* New York: Dover, 1958. (Reprint of the 1900 original.) The first five chapters (about 70 pages) of this charming old book contain a wealth of useful information about finite fields, especially about the number of solutions of polynomial equations. Much of the material in Chapter 11 of the present book was stolen from Dickson.

R. Lidl and H. Niederreiter, *Finite Fields.* Reading, Mass.: Addison-Wesley, 1983. (Encyclopedia of Mathematics and its Applications, v. 20). This is an excellent and huge book (800 pages) book devoted entirely to finite fields. Needless to say, it includes much material that the present book does not. I recommend it highly to those readers who wish to pursue the fascinating and sometimes difficult mathematics of finite fields.

W. W. Peterson and E. J. Weldon, Jr., *Error-Correcting Codes*, 2nd Ed. Cambridge, Mass.: MIT Press, 1972. This book was the first to be devoted to error-correcting codes, and is still notable for its extensive tables of irreducible polynomials over $GF(2)$.

Index

This book was designed and prepared in camera-ready form by the author. The composition was performed by TEX82 (version 1.1) using the "plain" macros plus a modest supplement. Phototypesetting was performed on an Autologic APS-μ5. Typefaces are from the Computer Modern family (the version dubbed "almost Computer Modern"). Both TEX and the Computer Modern family of typefaces are designs and implementations by Donald E. Knuth.